SpringerBriefs in Astronomy

More information about this series at http://www.springer.com/series/10090

SpringerBriefs in Astronomy

Series Editors

Martin Ratcliffe
Valley Center, Kansas, USA

Wolfgang Hillebrandt
MPI für Astrophysik, Garching, Germany

Michael Inglis
Long Island, New York, USA

David Weintraub
Vanderbilt University, Nashville, Tennessee, USA

More information about this series at http://www.springer.com/series/10090

Pamela Elizabeth Clark

Mercury's Interior, Surface, and Surrounding Environment

Latest Discoveries

Pamela Elizabeth Clark
Physics Department
Catholic University of America (IACS)
Washington DC
USA

ISSN 2191-9100 ISSN 2191-9119 (electronic)
SpringerBriefs in Astronomy
ISBN 978-1-4939-2243-7 ISBN 978-1-4939-2244-4 (eBook)
DOI 10.1007/978-1-4939-2244-4

Library of Congress Control Number: 2014953838

Springer New York Heidelberg Dordrecht London

Printed on acid-free paper

Springer is part of Springer Science+Business Media (www.springer.com)

Contents

About the Author

Pamela Clark grew up in New England. Inspired by President John Kennedy, she decided as a child to explore outer space. She thought, "If they can put a man on the Moon, they can put a woman (me) on Mars!" She obtained her BA from St. Joseph College. There, she had many opportunities to participate in laboratory research with Sr. Chlorophyll (Dr. Claire Markham) and Sr. Moon Rock (Dr. Mary Ellen Murphy) as well as to coordinate an NSF interdisciplinary undergraduate field research project. While obtaining her PhD in planetary geochemistry from the University of Maryland, she worked at NASA/GSFC outside of Washington, D. C., and the Astrogeology branch of the USGS in Flagstaff, Arizona, simulating, analyzing, correlating, and interpreting lunar X-ray spectra. She was a member of the group, led by Isidore Adler and Jack Trombka, that pioneered the use of orbital X-ray and gamma-ray spectrometers to determine the composition of planetary surfaces. She participated in the Flagstaff Lunar Data Consortium, the first attempt to create a common format database for all the remote sensing data from a planetary body.

After completing her PhD, Clark joined the technical staff at NASA/JPL, worked with the Goldstone Solar System Radar group, and expanded her remote sensing background to include radar, thermal, and near infrared studies of planetary surfaces with particular emphasis on the study of Mercury's surface. Clark organized a briefing team to promote a mission to Mercury, and for a while edited the Mercury Messenger newsletter.

Clark has published three books with Springer: *Dynamic Planet: Mercury in the Context of its Environment, Remote Sensing Tools for Exploration* (with Michael Rilee) and *Constant Scale Natural Boundary Mapping to Reveal Global and Cosmic Processes* (with Chuck Clark). She eventually returned to Goddard to work with the XGRS team on the NEAR mission to the asteroid Eros. She then became the science lead in a group initiated by Steve Curtis to develop new paradigms for the design of space missions and vehicles and to evaluate surface science scenarios, tools, technologies, and architectures for space missions to extreme environments, with particular emphasis on the Moon and Mars. Clark is currently an internationally recognized expert on the extension of the cubesat paradigm for high science return and low-cost exploration in deep space. She has done several stints in academia, including Murray State University in Kentucky, Albright College in Reading, Pennsylvania, and Catholic University of America in Washington, D. C. She has developed courses in analytical and environmental chemistry, geochemistry, physical geology, mineralogy, optics, planetary astronomy, remote sensing, and physics. Her goals include exploring under every rock to increase our sense of wonder about the Solar System.

Chapter 1
Background

In 2007, the author published the book *Dynamic Planet: Mercury in the Context of its Environment* to create a foundation, a snapshot, of then current knowledge and thinking about the planet, for the anticipated influx of new data from the NASA MESSENGER mission (Clark 2007). As expected MESSENGER, after three flybys and over two Earth years in orbit around Mercury, has provided the coverage necessary to get a 'global' view of the entire planet and its environment. Finally, observations of sufficient resolution and duration, particularly of the northern hemisphere, have created a firm basis for, in most cases, earlier weakly constrained hypotheses or, in some cases, revealed even more dynamic processes that challenge conventional views of the inner Solar System. For example, actual evidence for recent volcanic activity helps to substantiate the hypothesis that Mercury has an internal dynamo and a still partially molten interior to generate its global magnetic field.

A variety of short reviews were published after MESSENGER was in orbit (e.g., Kerr 2011; McKinnon 2012; Stevenson 2012; Witze 2011). Reactions to the new results from those interested in space exploration have varied. Some have emphasized the strangeness and uniqueness of the innermost planet, others how much the latest findings confirm how much Mercury is like the other terrestrial planets (Atkinson 2011; Grossman 2011; Spotts 2012; Wall 2012). Mercury's role as an end-member, the characteristics that make it more Earthlike than any of the other planets, as its presence in one of the most dynamic environments in the Solar System, was discussed before the MESSENGER mission and described thoroughly in *Dynamic Planet.* It didn't take the MESSENGER mission to reveal that Mercury is indeed *not* 'the boring planet' that many assumed four decades ago. MESSENGER was in fact conceived and designed about three decades after the *Mariner 10* flybys of Mercury provided the first in situ observations that were intriguing but of limited spatial and temporal coverage and resolution. Obviously, to justify a subsequent mission, Mercury's original image as 'boring' had to be completely displaced. Why was Mercury originally called boring (by Robert Strom), and how did it recover?

During and subsequent to the *Mariner 10* mission, work on the Viking missions and the first Mars landers was consuming most of NASA's resources for Solar System exploration at the USGS and JPL. The analysis of *Mariner 10* data was limited,

P. E. Clark, *Mercury's Interior, Surface, and Surrounding Environment*,
SpringerBriefs in Astronomy, DOI 10.1007/978-1-4939-2244-4_1

and the first set of Mercury geological maps remained incomplete for well over a decade. In the rush to get through the preliminary reports, the focus was on the most obvious and ubiquitous features at the available resolution—the craters. Of course, similar comments were made about Mars until well into the *Mariner 9* mission, when the global dust storms cleared. "Gee, all those bodies have craters, just like the Moon, so aren't they just like the Moon? Been there, done that."

A decade after *Mariner 10,* the results from that mission remained controversial. Earth-based astronomy programs continued to reveal further details of Mercury's surface topography, regolith, exosphere, and figure. As planetary scientists looked more carefully at the Mariner 10 observations, they observed a very complex surface, and hypothesized that several types of global-scale tectonic features (e.g., scarps, orthogonal ridges and troughs) revealed major tectonic episodes. In fact, a group of us organized, led, and participated in a workshop on Mercury 1987. The consensus was that we understood little about Mercury, and the planet had much to reveal about the origin of the solar system as well as current processes on our own planet. How could that tiny, supposedly 'dead' planet have a magnetic dynamo? What about all that evidence for resurfacing of an apparently volcanic nature, as opposed to the impact-generated deposits that had been postulated early on? How could Mercury be so dense, and yet yield so little evidence of iron in spectral observations? Mercury was observed to have a magnetosphere, making it the most Earth-like among the terrestrial planets, and that magnetosphere was suspected to be extremely dynamic due to its proximity to the Sun and its reduced size, and an exosphere, which Earth observations indicated varied as a function of Mercury's position in its annual cycle. Finally, the lack of geochemical observations from which bulk composition could be derived created great uncertainty as to not only Mercury's origin, but also the nature of formation processes in the early Solar System. In fact, far from being boring, Mercury was by that time considered to be a compelling and worthy target of exploration. The result was a series of briefings for the community and NASA headquarters, which resulted in a series of proposals for Mercury missions, as the planetary community began to see that it was essential that we fly another mission to Mercury that would provide more in-depth observations. MESSENGER was one of four proposals for Mercury missions under review simultaneously by the Discovery program. The major issue was how challenging a Mercury mission should be. Could we fly one with available technology? The MESSENGER mission proposed a highly elliptical, high-latitude periapsis orbit, and a Sun-facing sunshade that limited pointing capability of the spacecraft, all compromises to allow for survival in a high thermal Sun and surface flux environment.

Ground-based radar observations have contributed greatly to our understanding of Mercury. Prior to *Mariner 10,* Goldstein (1976) confirmed Mercury's 2:3 spin:orbit resonance. Ground-based radar observations provided topographic profiles of the equatorial region of Mercury, implying east/west asymmetry, as well as indications of water ice at the poles (Slade et al. 1992). Prior to MESSENGER, ground-based radar observations were used to precisely determine the spin axis and librations (Margot et al. 2007). Smith and coworkers (2012) combined these observations to determine the normalized moment of inertia, indicating a greater overall density and a core larger than previously assumed, requiring light-allowing elements, such as S and Si. Smith and coworkers (2012) also determined the ratio of moments of inertia for the outer shell and planet as a whole, indicating a thinner than previously assumed (400 km) rocky crust and mantle denser than Earth's, overlaying a molten Fe-Ni outer core, the basis for the magnetic dipole generating dynamo. Decoupling between outer and inner layers via a liquid layer introduces wobble, or libration (Stevenson 2012). Such an effect could induce a torque on Mercury's distorted (hot pole) shape, and as such contribute to major tectonic episodes evidence in global-scale features and volcanism during core expansion and contraction and large regional variations (northern hemisphere thinner) in crustal thickness. Such a thin crust would constrain the scale of tectonic activity and limit the dynamic range of topography (Zuber et al. 2015). Chemically reduced metallic melts of enstatite chondrite composition would help to explain the relatively high density of the mantle, as well as the production of a less dense iron sulfide layer between core and mantle, known as 'anti-crust.'

At this point, we'll give an overview of the mission, goals, and some of the more notable findings from MESSENGER. More detailed discussions of these will be discussed in later sections of this book.

1.1 MESSENGER Goals, Payload, and Design

Primary mission goals (Bedini et al. 2012) are described in Table 1.1 and include global mapping, determination of surface composition and by implication bulk composition, characterization of the exosphere, particle environment, magnetosphere,

Table 1.1 MESSENGER Goals

Surface mapping	B/W imaging	<250 m/pixel, 90 % coverage
	Multispectral imaging	<2 km/pixel, 90 % coverage
	Stereo from imaging	80 % coverage
	High resolution photometry	<<250 m/pixel, as able
Polar deposits	Composition of radar-reflective materials	Northern hemisphere
Dynamics/figure	Libration amplitude, gravitational field second degree components (derivation of moments of inertia), interior structure, obliquity	Improvement over previous measurements
Exosphere	Volatiles and neutrals, processes, sources, sinks	Improvement over *Mariner 10*
Particle environment	Ionized species, plasma processes, sources, sinks	Northern hemisphere
Magnetosphere	Structure and dynamics, plasma distribution, solar wind and exosphere interactions	Improvement over *Mariner 10*, primarily in northern hemisphere
Interior	Nature of magnetic field, interior structure and composition, origin	Improvement over *Mariner 10*, primarily in northern hemisphere

Table 1.2 MESSENGER Instruments

Payload	Measurement
MDIS	Wide-angle, narrow-angle cameras, surface imaging
GRNS	Fe, Ti, radioactive element composition to 10's cm depth, protons (ice) to 1 m
XRS	Major elements (Mg, Al, Si, S, Ca, Ti) to 10's microns depth
MLA	Topography, planetary shape
MASCS	Exosphere species, surface mineralogical and volatile measurements
EPPS	Charged particle and plasma environment
MAG	Planetary magnetic field

and magnetic field, as well as interactions between surface, exospheric, and magnetospheric particles and the surrounding environment.

The payload used to acquire measurements needed to achieve these goals include the Mercury Dual Imaging System (MDIS) (Hawkins et al. 2007), the Gamma-ray and Neutron (GRNS) (Goldsten et al. 2007), and X-ray (XRS) spectrometers (Schlemm et al. 2007), the Magnetometer (MAG) (Anderson et al. 2007), Laser Altimeter (MLA) (Cavanaugh et al. 2007), Atmospheric and Surface Composition Spectrometer (MASCS), consisting of the Ultraviolet/Visible (UVVS) and Visible/Infrared (VIRS) spectrometer (McClintock and Lankton 2007), as well as the Energetic Particle and Plasma Spectrometer (EPPS) (Andrews et al. 2007) (Table 1.2). To minimize complexity, only MDIS has 'moving parts'—a 1° of freedom gimbal to facilitate image capture. Thus, instruments are not nadir pointed, and the spacecraft must be turned to downlink data and altitude actively managed to 'point' the instruments.

Launched in August 2004, MESSENGER had a six and a half year cruise, during which it completed three flybys of Mercury (January 2008, October 2008, September 2009), before achieving orbital insertion in March 2011. Even with multiple gravity-assist flybys, more than half of the total launch mass (1200 kg) was propellant required for orbital insertion near the Sun's gravity. To minimize the mass of onboard fuel required, MESSENGER used a solar sailing technique developed and tested by *Mariner 10* (Clark 2007; O'Shaughnessy et al. 2011), using solar radiation pressure on tilted solar panels to control spacecraft momentum passively. Mercury's environment is thermally challenging and requires special considerations in spacecraft and mission design. A specially designed highly reflective sunshade, consisting of layers of kapton and nextel ceramic cloth on a titanium frame, is deployed to protect the spacecraft from direct sunlight. In addition, the spacecraft 12-h orbit is highly elliptical and highly inclined, with high northern latitude periapsis, and during 'hot pole' periods the equator crossing is designed to be near the terminator. Although they impact achievement of science goals, especially in terms of limiting coverage in the southern hemisphere and thus global coverage, these strategies are crucial for maintaining the spacecraft within operational limits. During the course of one Earth year (two Mercury solar days), a specific location on the planet can be observed twice with the same viewing geometry.

1.2 MESSENGER Highlights

Some of the most significant results from MESSENGER (Bedini et al. 2012; Kerr 2012; McKinnon et al. 2012; Stevenson 2012), which will be discussed in great detail on the subsequent chapters, include:

- Completion of the global mosaic, finally revealing the hemisphere not imaged by *Mariner 10*.
- Evidence of two major episodes of volcanism, including extensive flooded lowland northern plains, as well as younger/older globally distributed volcanic terrain, with subtle variations in style within terrains.
- Discovery of 'hollow' depressions in bright deposits within impact craters, which appear to be shaped by relatively recent events.
- First direct measurements of surface composition, indicating a surface relatively low in iron silicate as anticipated from ground-based spectral reflectance observations, but rich in Mg and, unexpectedly, in volatiles, especially sulfur and potassium, yielding a K/Th ratio comparable to other terrestrial planets.
- Mercury's bulk composition, implied from direct chemical measurements as well as gravitational and figure parameters, eliminates most planetary formation models, which require loss of volatiles. Low iron silicate (and an iron-rich core required for the dipole) indicate reducing conditions in the inner nebula, unlike those of the other terrestrial planets. By implication, the closest match for bulk composition is enstatite chondrite, rich in Mg, with more sulfur and metallic iron than other terrestrial planet compositions.
- Global altimetry, magnetic, and gravitational fields, determined roughly in the southern hemisphere and at higher resolution in the northern hemisphere, indicating a thinner northern crust and supporting model of thinner crust, denser mantle, and larger core than previously assumed, as well as solid iron sulfide-rich layer between mantle and crust.
- Confirmation of the magnetic dipole field and, based on gravitational parameters, of a partially molten liquid iron core (dynamo).
- An offset from the geographic equator of 20 % of Mercury's radius in the global magnetic field, with implications for no offset in the core but magnetosphere structure and particle interactions with Mercury's surface, implying the southern poles especially could be an important source of the ionized exosphere.
- Measurements indicating no high radiation 'belts' such as the Van Allen Belt, but bursts of energetic electrons across a wide range of latitudes and times of day and efficient acceleration processes in Mercury's magnetosphere operating on time scales an order of magnitude faster than analogous processes in Earth's magnetosphere.

References

Anderson, B., Acuna, M., Lohr, D., Scheifele, J., Raval, A., Korth, H., Slavin, J.: The magnetometer instrument on MESSENGER. Sp. Sci. Rev. **131**, 417–450 (2007)

Andrews, G., Zurbuchen, T., Mauk, B., Malcom, H., Fisk, L., Gloeckler, G., Ho, G., Kelley, J., Koehn, P., LeFevere, T., Livi, S., Lundgren, R., Raines, J.: The energetic particle and plasma spectrometer instrument on the MESSENGER spacecraft. Sp. Sci. Rev. **131**, 523–556 (2007)

Atkinson, N.: MESSENGER's first image from Orbit of Mercury, Universe Today, March 29, 2011. http://www.universetoday.com/84488/messengers-first-image-from-orbit-of-mercury/ (2011). Accessed 6 Jan 2014

Bedini, P., Solomon, S., Finnegan, E., Calloway, A., Ensor, S., McNutt, R., Anderson, B., Prockter, L.: MESSENGER at mercury: a mid-term report. Acta. Astronaut. **81**, 369–379 (2012)

Cavanaugh, J., Smith, J., Sun, X., Bartels, A., Ramos-Izquierdo, L., Krebs, D., McGarry, J., Trunzo, R., Novo-Gradac, A., Britt, J., Karsh, J., Katz, R., Lukemire, A., Szymkiewicz, R., Berry, D., Swinski, J., Neumann, G., Zuber, M., Smith, D.: The mercury laser altimeter instrument for the MESSENGER mission. Sp. Sci. Rev. **131**, 451–479 (2007)

Clark, P.E.: Dynamic Planet: Mercury in the context of its Environment, p. 229. Springer (2007)

Goldstein, R.: Radar observations of mercury. Astrophys. J. **210**, 250 (1976)

Goldsten, J., Rhodes, E., Boynton, W., Feldman, W., Lawrence, D., Trombka, J., Smith, D., Evans, L., White, J., Madden, N., Berg, P., Murphy, G., Gurnee, R., Strohbehn, K., Williams, B., Schaefer, E., Monaco, C., Cork, C., Eckels, J., Miller, W., Burks, M., Hagler, L., Deteresa, S., Witte, M.: The MESSENGER gamma-ray and neutron spectrometer. Sp. Sci. Rev. **131**, 339–338 (2007)

Grossman, L.: Brilliant new images of Mercury from first year of orbit, Wired Science, 17 June 11. http://www.wired.co.uk/news/archive/2011-06/17/mercury-new-pictures (2011). Accessed 6 Jan 2014

Hawkins, S.E., Boldt, J., Darlington, E., Espiritu, R., Gold, R., Gotwols, B., Grey, M., Hash, C., Hayes, J., Jaskulek, S., Kardian, Jr, C., Keller, M., Malaret, E., Murchie, S., Murphy, P., Peacock, K., Prockter, L., Reiter, R., Robinson, M., Schaefer, E., Shelton, R., Sterner, R., Taylor, H., Watters, T., Williams, B.: The mercury dual Imaging system on the MESSENGER spacecraft. Sp. Sci. Rev. **131**, 247–338 (2007)

Kerr, R.: mercury looking less exotic, more a member of the family. Science. **333**, 2011 (1812)

Margot, J.L., Peale, S., Jurgens, R., Slade, M., Holin, L.: Large longitude libration of mercury reveals a molten core. Science. **316**, 710 (2007)

McClintock, W., Lankton, M.: The mercury atmospheric and surface composition spectrometer for the MESSENGER mission. Sp. Sci. Rev. **131**, 481–521 (2007)

McKinnon, W.B.: The strangest terrestrial planet. Science. **336**, 162–163 (2012)

O'Shaughnessy, J., McAdams, V., Bedini, P., Calloway, A., Williams, K.: MESSENGER's use of solar sailing for cost and risk reduction. Proc 9th Low-Cost Planetary Missions Conference, IAA, Laurel, MD, June 21–23 (2011).

Schlemm, C., Starr, R., Ho, G., Bechtold, K., Benedict, S., Boldt, J., Boynton, W., Bradley, W., Freeman, M., Gold, R., Goldsten, J., Hayes, J., Jaskulek, S., Rossano, E., Rumpf, R., Schaefer, E., Strohbehn, K., Shelton, R., Thompson, R., Trombka, J., Williams, B.: The X-ray spectrometer on the MESSENGER spacecraft. Sp. Sci. Rev. **131**, 393–415 (2007)

Slade, M., Butler, B., Muhleman, D.: Mercury radar imaging: evidence for polar ice. Science. **258**, 635–640 (1992)

Smith, D.E., Zuber, M., Phillips, R., Solomon, S., Hauck, S., Lemoine, F., Mazarico, E., Neumann, G., Peale, S., Margot, J.-L., Johnson, C., Torrence, M., Perry, M., Rowlands, D., Goossens, S., Head, J., Taylor, A.: Gravity field and internal structure of mercury from MESSENGER. Science. **336**, 214 (2012)

Spotts, P.: Strange features on Mercury upend thinking about 'first rock from sun', Christian Science Monitor, March 21, 2012. http://www.csmonitor.com/Science/2012/0321/Strange-features-on-Mercury-upend-thinking-about-first-rock-from-sun (2012). Accessed 6 Jan 14

Stevenson, D.: Mercury's mysteries start to unfold. Nature. **485**, 52–53 (2012)

Wall, M.: Mercury Surprises: Tiny Planet has strange innards and active past, Space.com, 3/21/12. http://www.space.com/14978-mercury-discoveries-messenger-spacecraft.html (2012). Accessed 6 Jan 2014

Witze, A.: Mercury revealed. Nat. Geosci. **5**, 303 (2012)

Zuber, M., Smith, D., Phillips, R., Solomon, S., Neumann, G., Hauck, S., Peale, S., Barnouin, O., Head, J., Johnson, C., Lemoine, F., Mazarico, E., Sun, X., Torrence, M., Freed, A., Klimczak, C., Margot, J.-L., Oberst, J., Perry, M., McNutt, R., Balcerski, J., Michel, N., Talpe, M., Yan, D.: Topography of the Northern hemisphere of mercury from MESSENGER laser altimetry. Science. **336**, 217 (2012)

Chapter 2
Mercury's Interior

2.1 Mercury's Composition and Its Significance

MESSENGER has obtained the first direct compositional measurements of Mercury, measurements that are essential as a basis for constraining the planet's bulk composition and origin. The relatively unconstrained composition of Mercury allowed for a wide range of formation models (Fig. 2.1, Table 2.1) for not only the planet but also the Solar System (Clark 2007). Now that range is considerably reduced in a somewhat surprising direction, implying that enstatite chondrite is the closest 'match' among proposed models.

Table 2.2 summarizes our current assessment of Mercury's bulk and major terrain composition based on MESSENGER XRS, GRS, and neutron spectrometer measurements of elemental abundances (Weider et al. 2012, 2014; Nittler et al. 2011; Evans et al. 2012; Peplowski et al. 2012a, b; Stockstill-Cahill et al. 2012). Elemental and mineral abundance maps and regional averages derived from these instruments are described below.

Some of the most compelling findings are:

1. Relatively high Mg (magnesium) and Mg silicates, but low to moderate total iron and low Fe (iron) silicates combined with a large, dense iron-rich core implied by both density measurements and the presence of a confirmed dynamo.
2. Low Al (aluminum) but moderate Ca (calcium) correlated with Mg abundance and found along with Mg in silicates and sulfides.
3. Elevated volatile abundances including S (sulfur), Na (sodium), and K (potassium) on the surface and in the atmosphere, as well as the presence of ices and sulfides.

Enstatite chondrites, a meteorite class represented by the Abee Meteorite postulated to be an impact breccia from Mercury (Griffin et al. 1992), are the most reduced chondrites, characterized by high metallic iron (20–30 %) and low abundances of iron-bearing silicate (<5 %). Chondrules contain very magnesian enstatite. However, the inner Solar System has no clear asteroid 'parent' that can be linked to Mercury.

P. E. Clark, *Mercury's Interior, Surface, and Surrounding Environment*,
SpringerBriefs in Astronomy, DOI 10.1007/978-1-4939-2244-4_2

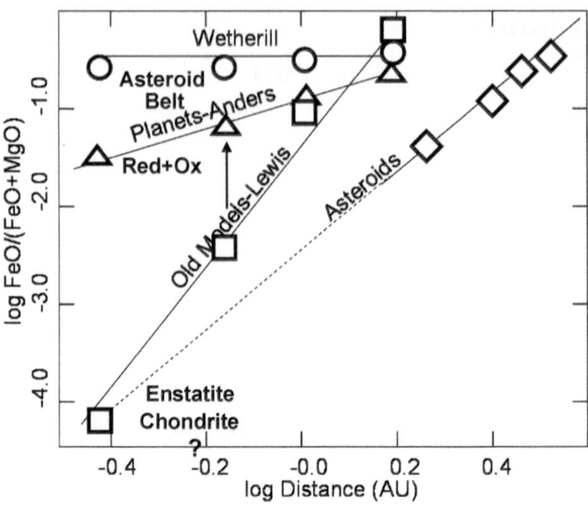

Fig. 2.1 Wide range of formation models possible with unconstrained Mercury composition. (Courtesy of Ken Goettel 1988, unpublished figure)

Table 2.1 Planetary formation model implications for bulk composition

Planetary formation model	Elements						
	Mg	Fe	Si	Na (alk)	Al (refr)	Th (resid)	S (vol)
(EQ) equilibrium condensation	H	LL	M	L	L	H	L
(SA) selective accretion	ML	MH	L	H	MH	H	M
(TV) t tauri vaporization	M	ML	L	L	HH	HH	L
(GI) giant impact	ML	MH	L	M	L	L	L
(RO) reduced/ oxidized	L	H	M	L	L	L	M
(EC) enstatite chondrite	H	M	H	H	L	L	H

H high, *M* moderate, *L* low

EQ Lewis Heliocentric distance dependent temperature, *SA* selection of metals versus silicates due to aerodynamic/mechanical properties, *TV* removal of large fraction of silicate mantle and volatiles due to T Tauri event, *GI* removal of large fraction of silicate mantle and volatiles due to giant impact, *RO* Anders reduced (metal enriched) followed by provenance-dependent incorporation of oxidized (silicate enriched) components, *EC* reduced chondritic composition at Mercury provenance

The compositions of major terrains, indicative of major episodes of geochemical differentiation, are shown in Table 2.2. The older, intercrater plains/highly cratered terrain (IcP/HCT) is richer in Mg and lower in Fe, implying a higher temperature melt, as indicated by spatially resolved measurements of Mg/Si, Al/Si, S/Si, and Ca/Si for the northern volcanic plains. Early volcanism involving deposition of this

Table 2.2 XRS- and GRS-derived bulk and major terrain composition

Abundance	Source	Bulk	ICP/HCT	NVP	CB
XRS-derived					
Mg/Si	Weider et al. 2012		0.57	0.34	
S/Si	Weider et al. 2012		0.09	0.06	
Ca/Si	Weider et al. 2012		0.19	0.15	
Fe/Si	Weider et al. 2014	0.06			
Al/Si	Peplowski et al. 2012b		0.22	0.26	
Mg	Weider et al. 2012		14.4	8.6	
Fe	Weider et al. 2014	1.5			
Ca	Weider et al. 2012		4.9	3.7	
S	Weider et al. 2012		2.3	1.5	
Al	Weider et al. 2012		5.4	6.6	
GRS-derived					
S/Si	Evans et al. 2012	0.092			
Ca/Si	Evans et al. 2012	0.24			
Fe/Si	Evans et al. 2012	0.077			
Al/Si	Peplowski et al. 2012b	0.25			
Na/Si	Evans et al. 2012	0.12			
Si	Evans et al. 2012	24.6			
Mg	Peplowski et al. 2012b	12.2			
Fe	Evans et al. 2012	1.9			
Ca	Evans et al. 2012	5.9			
S	Evans et al. 2012	2.3			
Al	Peplowski et al. 2012b	5.7			
Na	Evans et al. 2012	2.9			
O	Peplowski et al. 2012b	48.9			
K	Peplowski et al. 2012a	1200 ppm	952 ppm	1786 ppm	754 ppm
Th	Peplowski et al. 2012a	0.16 ppm	0.168 ppm	0.142 ppm	0.54 ppm
U	Peplowski et al. 2012a	90 ppb			
Ti	Rhodes et al. 2011	1.2			
Si	Rhodes et al. 2011	18			

% abundance and abundance ratios by weight

material was followed by two episodes of smooth plains (SP) formation, which apparently overlapped with northern plains formation. Fe progressively substituted for Mg as the interior cooled during a successively lower temperature melting process. The observed relatively high Na abundance is consistent with the presence of Na-bearing plagioclase feldspar inferred from ground-based spectral measurements (Sprague and Massey 2007).

2.2 Geochemical Differentiation

Either Mercury accreted from source material similar to enstatite chondrites but with higher bulk Fe or the enstatite chondrite parent body and Mercury accreted independently from material of similar composition, in terms of bulk abundances, mineralogy, and isotope compositions. In either case, a high degree of partial melting of a very reduced source and crystallization of basalts would have occurred subsequently. The composition of Mercury surface material is apparently more similar to basaltic melt derived from an enstatite chondrite-like source region rather than to residual silicates such as aubrites, particularly in regard to the widespread presence of sulfides (Weider et al. 2012).

High surface S combined with low oxygen fugacity implies accretion from reduced materials such as enstatite chondrites, which might also have been subjected to loss and depletion of iron sulfide resulting from evaporation of troilite in the inner nebula (Zolotov et al. 2013). Models of the geochemical differentiation process (Zolotov et al. 2013) based on these findings indicate that calcium and magnesium sulfides would be common in volcanic rocks, and oxygen fugacities would be low in parent magmas. Zolotov and coworkers (2013) evaluated resulting oxygen and sulfur fugacities as well as resulting silicate/sulfide equilibria using empirical models for silicate and metallurgical slag formation. The predicted range of oxygen fugacities (IW-4.5–7.3@1700–1800 K) (Zolotov et al. 2013) has implications for the differentiation process. Resulting values of Fe in silicate correspond to 0.03–0.79 weight percent FeO, below the detection limit of reflectance measurements, are consistent with reducing conditions, lower oxygen fugacities, and partial melt of enstatite chondrite. These conditions imply the presence of iron in sulfides, also suggested by the correlation between Fe and S. Low oxygen fugacities and reducing conditions, combined with the presence of S constrain oxidation states, increase sulfur solubility and induce more 'chalcophile' behavior (bonding with S) among typically 'lithophile' (bonding with Si) elements, including Mg, Ca, Ti, Na, and K, as well as U and Th, and thus affect metal/silicate partitioning (Keil 1968).

Abundances of major rock-forming elements (Mg, Al, Ca, Si) also indicate a high degree of partial melting implying substantial heat production in the mantle (Peplowski et al. 2011) (Table 2.2). Extensive early volcanism followed by episodic volcanism supports core formation-induced crustal expansion, followed by slow cooling, and finally core partial solidification induced crustal contraction. Implied bulk compositions are most similar to Mars meteorites, depleted in incompatible elements (Peplowski et al. 2011).

Peplowski and coworkers (2011, 2012a) GRS-derived average surface abundances for radioactive elements K, Th, and U (Table 2.2) are consistent with chondritic source from high temperature nebular condensates but, based on the K/Th ratio, with considerably greater volatile abundances than conventionally associated chondritic material. Observed U/Th ratios are slightly less than inner planetary and chondritic averages (Peplowski et al. 2011), indicating some separation of U (uranium), thus somewhat lower oxygen fugacity and reducing conditions, because U and Th (thorium) are not separated under more oxidizing conditions. However, U and

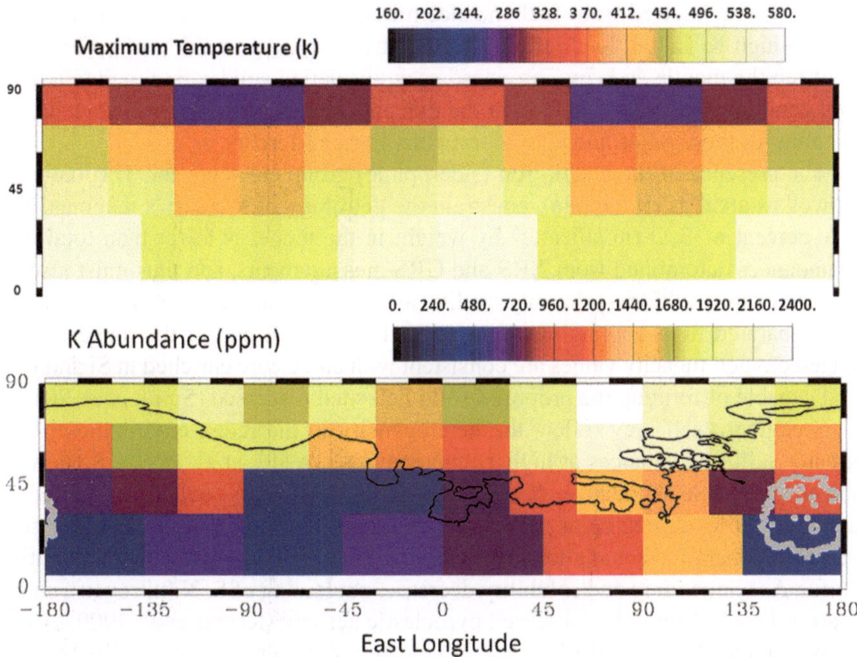

Fig. 2.2 Comparison of maximum temperature (*top*) and K abundance (*bottom*) from 0 to 90 degrees latitude (*vertical axis*) and all longitudes (*horizontal axis*). (Peplowski et al. JGR Planets 2012a, Fig. 11b (*top*) and 8a (*bottom*), © 2012 American Geophysical Union)

Th abundances on Mercury are comparable to terrestrial oceanic basalts (Peplowski et al. 2011), limiting the extent to which reducing conditions and sequestration of these elements with heavy element sulfides occur in the core. Thus, although K/Th ratio is higher than inner planetary and chondritic averages, that ratio can't be explained by extensive core sequestration of Th but is more likely to result from thermal redistribution of K near the surface. In addition, models based on these bulk abundances indicate not only considerable internal heating but for a period four times longer than previously thought, with not enough heat production from incompatible radioactive elements (U, Th) to drive core convection (Peplowski et al. 2011, 2012a). K, Th, and U abundances are thus consistent with a model for widespread volcanism following the late heavy bombardment, gradually declining internal heat production and with less extensive episodes of volcanic activity subsequently (Fig. 2.2). McCubbin and coworkers (2012) also interpreted GRS and XRS-derived K, Th, U, and S abundance values to be indicative of a higher reducing environment with very low oxygen fugacity and higher volatile abundances than previously assumed. They pointed out that an important indicator of the volatile depletion during formation is the ratio of very large ion lithophiles (LIL) that are relatively volatile (e.g., K) to those that are more refractory (U, Th, rare earths). U and Th partitioning into the metal rich core and the sulfide oldhamite ([CaMgFe]S), which can be inferred from S abundance measurements, is not enough to account for the high K/Th and K/U ratios. On the other hand, the addition of a FeS layer between the mantle and core, proposed as an insulating solid cap to maintain thermal conditions necessary to sustain the dynamo that will be discussed soon, could account for these ratios and allow for a volatile-rich Mercury.

The presence of the nickel-iron (siderophile) group elements and Ti (titanium), as well as alkalkis (K and Na), constrain the petrology of Mercury's magmas. The low percent of FeO (in silicates) by weight in the model is lower than total iron abundances determined from XRS and GRS measurements, and thus must also be present as iron sulfides, primarily troilite, and metal, which could have crystalized and separated from cooling magmas. All but the lowest part of the range of predicted oxygen fugacity values are consistent with an Fe core enriched in Si and/or S and capable of forming the proposed solid FeS shell discussed (Smith et al. 2012).

S content itself may reflect the heterogeneity in the redox environment, e.g., greater sulfur abundances at higher temperatures (Weider et al. 2012). Somewhat lower Mg and So abundances of the northern volcanic plains could be explained by a somewhat lower degree of partial melting than the oldest terrains. A lower subsurface temperature could allow partial oxidation of sulfide and formation of native sulfur. As temperature and solubility decrease, outgassing of S, N (nitrogen), and C (carbon) could lead to the observed pyroclastic activity (Kerber et al. 2009, 2011). However, the observed effusive style of volcanic activity for many melts apparently indicate a minor role for outgassing resulting from assimilation of crustal S (Head et al. 2011).

Despite the presence of sulfides and immiscibility of silicate and sulfide melts, the predominance of silicates combined with their far higher temperatures of crystallization would not be consistent with large-scale separation and flotation of less dense Ca and Mg sulfides on the silicate magmas (Zolotov et al. 2013; McCoy et al. 1999). Instead these sulfides, combined with Fe, were embedded in the accreting material, and distributed as late stage crystallization products in erupting magmas. Dense Fe-S rich melts, on the other hand, would sink, ultimately forming a mantle sublayer (Smith et al. 2012).

Ebel and Alexander (2011) have pointed out that solids likely to survive in the accretion zone of Mercury would be similar to highly unequilibrated anhydrous interstellar organic and presolar grains with porous interplanetary dust particles of chondritic composition (C-IDP) as follows. High temperature reactions with these compositions produce condensates (solid plus liquid assemblages in equilibirium with vapor) with greater than chondritic Fe/Si ratios that are about half of the Fe/Si estimated for the bulk composition of Mercury. Stable minerals include FeO-poor enstatite chondrites, with S behaving as a refractory element at lower temperatures. Local compositional gradients in volatiles of the accreting disk, due to 'cold finger' effects (Stevenson and Lunine 1988), combined with an efficient metal/silicate fractionation process, could create enstatite parent body compositions and explain Mercury's anomolous (high S, very low FeO, high metallic Fe) composition. The condensing Mg-rich olivine in C-IDP systems removes oxygen, progressively decreasing oxygen fugacity and FeO content They reduce even more as the temperature decreases, in direct contrast to hydrous silicate-rich CI systems. Eventually enstatite predominates over olivine, creating the basis for Mercury's low Fe silicate composition with a metallic iron core. In Mercury's neighborhood, unlike the neighborhoods of other terrestrial planets, the presence of C-IDP dust resulted in sulfur in refractory rather than vapor phases (Ebel and Alexander 2011). Thus, Ca occurs with S rather than Al, as it does in oxidized systems. Si rather than Fe is more volatile over a broader temperature range, resulting in a high bulk Fe/Si ratio. If this model is correct, Mercury should be relatively organic rich and hydrogen poor, reflecting the I-CDP composition.

2.3 Surface Composition and Its Implication for Interior Processes

The distributions of GRS measurement-derived elements (Table 2.2) are dissimilar (Peplowski et al. 2012a, b). Th, Si, and O vary little. K is anti-correlated with maximum surface temperatures, varying from 300 to 2400 ppm in the northern hemisphere, and thus apparently controlled thermally, lost to the exosphere where it has been observed (Ref). Its higher value for the northern volcanic plains would be consistent with their location around a 'cold pole' described earlier (Fig. 2.2). The decrease in K/Th is consistent with thermally induced diffusion rates in alkali-bearing feldspars. Another clue substantiating K loss is the pattern of minimum Na/K ratio at the subsolar point, which is consistent with a photo sputtering rate for K twice that of Na and K's far longer lifetime before ionization. Exospheric Na is 250 times greater on Mercury than on the Moon, whereas exospheric K is only 93 times greater on Mercury, although K is far more enhanced at warm longitudes.

On the other hand, K/Th ratios could show a different relationship for different episodes of volcanism as they do on the Moon (Fig. 2.3). The highest K and Th on the Moon are associated with the oldest volcanic basalts, with lower abundances and considerable variation found in earlier crustal rocks. However, temperature differences as opposed to compositional differences are likely to account for most of the variations in surface K on Mercury.

XRS and GRS derived Fe/Si and Ti/Si ratios (Table 2.3) are consistent with values estimated from the neutron spectrometer. GRS Si abundance estimates are consistent with the earlier observations from ground-based IR measurements that Mercury rocks have slightly less Si than lunar rocks (Sprague and Massey 2007). However, little Fe and virtually no Ti are observed, contrary to earlier interpretations of color observations indicative of high Ti basalts. Thus, FeTi oxide abundances would be insufficient to account for neutron absorption inferred from the neutron spectrometer. Because additional metallic Fe in nanophase iron or elsewhere isn't compatible with the presence of FeO-rich ilmenite, Riner and coworkers (2010) had proposed rhombic oxide $MgTiO_3$ (egelkelite), which could be produced by partial melting of metal-rich enstatite chondrite (Brown and Elkins-Tanton 2009) as an alternative, a possibility that is still excluded by the extremely low Ti abundances.

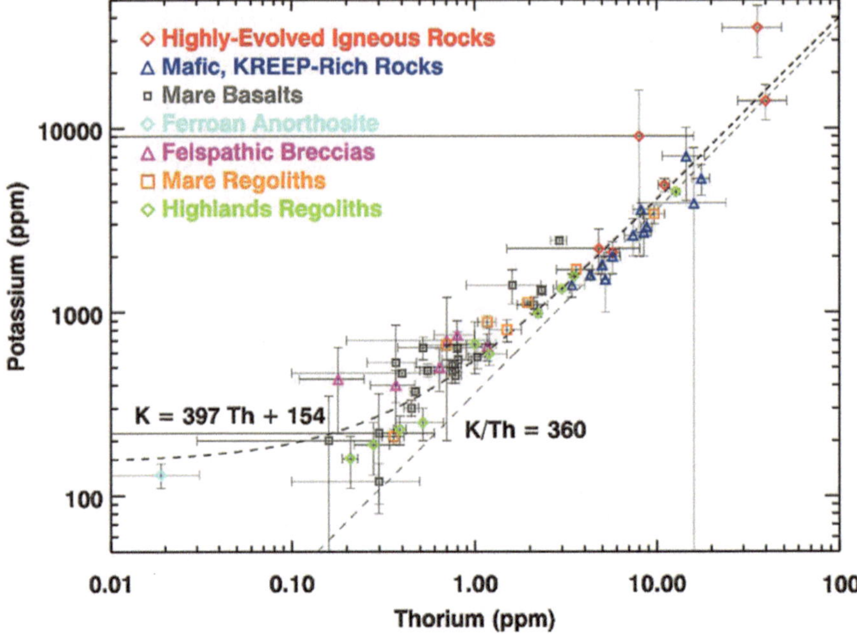

Fig. 2.3 Relationship between lunar K and Th indicating K retention relative to Th for lower Th rocks. Mercury has a low Thorium bulk abundance. Peplowski et al., JGR Planets, 2012, Figure 10, ©2012 American Geophysical Union.

Table 2.3 Dynamic figure parameters

Parameter	Value	Error	Source
Radius	2439 km	0.69	Oberst et al. 2011
Crust and mantle depth	410 km		Hauck et al. 2013
Depth to liquid core	420 km		Hauck et al. 2013
Spin axis	2.04 arc min	0.08 arc min	Margot et al. 2012
Forced libration amplitude	38.5 arc sec	1 arc sec	Margot et al. 2012
Forced libration displacement	450 m	10 m	Margot et al. 2012
C/MR^2	0.346	0.014	Margot et al. 2012
C_M/C	0.431	0.025	Margot et al. 2012
C_{20} dynamic oblateness	2.24×10^{-5}	0.01×10^{-5}	Smith et al. 2012
C_{22} dynamic ellipticity	1.253×10^{-5}	0.01×10^{-5}	Smith et al. 2012

Mercury has a distinctive and anomalous surface composition. Crustal iron is correlated with magnetization contrast in all the terrestrial planets except Mercury (Purucker et al. 2009). Weider and coworkers (2014) mapped average XRS Fe/Si ratio measurements taken during 55 solar flares (Fig. 2.4) and an estimated bulk iron abundance of 1.5 % from these measurements. Most of the measurements

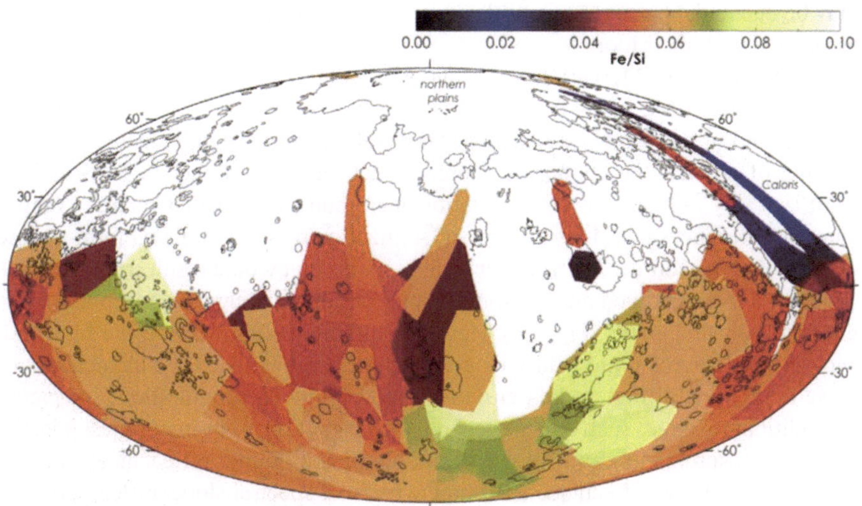

Fig. 2.4 Map of XRS-derived Fe/Si measurements taken during times of solar flares. Reprinted from Icarus, 235, Weider et al, Figure 10a, ©2014, with permission from Elsevier.

included in the study were taken from the southern hemisphere when the spacecraft was further away from the planet and areas of coverage were larger. Thus, we have little coverage for the more extensively studied volcanic deposits in the northern hemisphere. Attempts were made to correct measurements for differences in phase angle. Corrected measurements, with a 10 % intrinsic error, vary by two standard deviations form the average across the southern hemisphere. Significantly, larger variations in Fe/Si are observed for more spatially resolved, smaller footprint areas. Largest Fe/Si ratios, potentially indicating larger iron metal and iron sulfide deposits, are found in the southern hemisphere associated with 'blue' areas.

The relatively high Mg/Si and low Al/Si and Ca/Si ratios clearly indicate Mercury does not have a lunarlike crust, and the sulfur abundance makes it substantially more chondritic than the Moon and terrestrial planets (Table 2.2). These data, combined with the low surface iron and iron silicate abundances, suggest highly reduced parent material such as enstatite chondrite or anhydrous dust particles, although additional metal/silicate fractionation in the solar nebular may be necessary to explain Mercury's bulk composition. Surface composition ranges from ultramafic to basaltic and most closely resembles terrestrial komatiites, which could have formed from a high degree of partial melting of the proposed parent material. These measurements and proposed parent material are consistent with the presence of Mg- and Ca-rich sulfides as opposed to FeS as the dominant form of sulfur in the crust. The presence of sulfur would support the development of pyroclastics during volcanic activity. Low Ti and Fe do not support the conventional model for space weathering based on the darkening of opaque oxides, although high surface temperatures and aggressive particle bombardment could support formation of agglutinates. Low reflectance, particularly associated with areas of the highest Mg/Si, Ca/Si, and S/Si, may be due to the presence of sulfides as opposed to conventional space weathering associated with the Moon.

Major terrains of Mercury have distinctive signatures (Weider et al. 2012, 2014) as follows. High Mg mafic minerals, plagioclase feldspar, and smaller amounts of Ca, Mg, and/or Fe sulfides are the major minerals reflecting a high degree of partial melting of an enstatite chondritic source. Older terrain is higher in XRS-derived Mg/Si, S/Si, Ca/Si and lower Al/Si ratios than younger, smoother plains. Compositional differences could result from crystallization of smooth plains from a more evolved magma source. The younger Caloris plains that would result from crystallization of a more evolved magma source can be readily distinguished from the older northern volcanic plains (Fig. 2.4). However, although substitution of Fe for Mg in Mg-depleted source regions in younger volcanic units is likely, Fe abundances of these units are still higher than enstatite chondrite melts should allow, indicating an additional source, such as an exogenous contribution from meteorites. Visible/near IR measurements can distinguish smooth plains (higher reflectance, from intermediate and low reflectance units on the basis of spectral slope, reflectivity, and morphology) (Denevi et al. 2009). The two clusters of the oldest terrain (IcP-HCT) have higher Mg/Si, S/Si, Ca/Si, and lower Al/Si than the northern plains. One cluster has somewhat more moderate Mg/Si and Al/Si. The oldest terrain is most compositionally analogous to terrestrial ultramafic komatiites, but with a larger range of composition than seen in terrestrial midoceanic ridge basalts.

Mg and Fe compositions are consistent with Mg-rich pyroxene (enstatite) and olivine (forsterite), and Ca/Si and S/Si ratios are most consistent with the sulfide oldhamite (Weider et al. 2012, 2014), which has been identified by Sprague and coworkers (1995) from ground-based IR observations and is found in enstatite chondrites and achondrites. Most of the Ca is probably in plagioclase feldspar, but the lack of correlation with Al indicates that Ca must be found in minerals other than calcic plagioclase as well, including sulfides and non-sulfides such as diopside, which has also been identified by Sprague and coworkers (2002). Ca may also be found in Ca-rich pyroxene, but that would not account for the higher Al/Si ratios for the younger northern volcanic plains, indicating that some of the Al must be found in Na- or K -ich feldspars, as inferred from GRS observations (Peplowski et al. 2012a, b), although variations in K are related to surface heating.

The bimodal composition of IcP/HCT terrain could suggest that the crust is heterogeneous (Weider et al. 2014). The two distinct lithologies are ultramafic with enstatite and some plagioclase and oldhamite and basalt with some sulfide content (more like the northern plains material). Lower Mg of the younger material suggests derivation from more differentiated and cooler magma from which higher Mg mafics have already crystallized, producing a composition more like mid-oceanic ridge basalts (MORB). If the younger northern plains and Caloris Basin smooth plains resulted from remelting of the older basalt source region, the trends in composition would parallel komatiite-MORB trends, but this is not the case (Fig. 2.5). The more likely scenario is that the two terrains originated from different portions of the mantle, the earlier at higher temperature and greater partial melting, and the latter at lower temperature due to slow cooling and thus a lesser degree of partial melting.

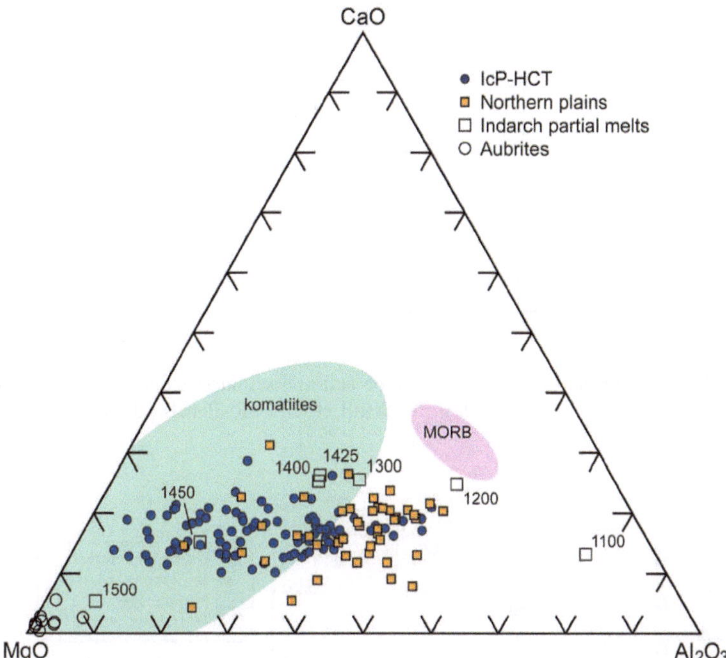

Fig. 2.5 MgO-Al2O3-CaO ternary diagram illustrating relationship between older IcP-HCT and northern plains volcanic units, enstatite chondrite partial melt (Indarch), aubrite meteorites, and terrestrial ultramafic (komatiites) and mafic (MORB) rocks.(Weider et al. JGR Planets 2012, Fig. 6, © 2012 American Geophysical Union)

Mercury mineral compositions and abundances have been derived using petrological modeling of Mg-rich magmatism under reducing conditions assumed on the basis of bulk compositional estimates for Mercury (Stockstill-Cahill et al. 2012). The MELTS model, typically used for terrestrial applications, was modified for Mercury conditions by assuming a lower initial iron content and removing sulfur, on the basis that it would form sulfides; corresponding mole fractions of Mg and Ca were removed. The resulting models indicated that the closest analog for Mercury rocks would be magnesian basalt composed of Mg-rich orthopyroxene and plagioclase (Fig. 2.6).

2.4 Thermal Considerations and Implications for the Interior

A variety of measurements provided by MESSENGER have yielded constraints for the thermal evolution of Mercury (Michel et al. 2013). Surface Mg and S abundances are consistent with high interior temperatures and a high degree of partial melting

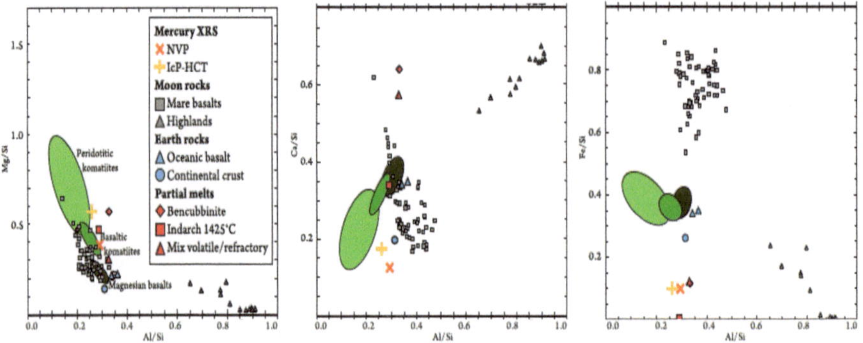

Fig. 2.6 MELTS models of Mercury rock type and major minerals derived from relationships between Mg/Si, Fe/Si, Ca/Si, and Al/Si. Stockstill et al., JGR Planets, 2012, Figure 2, © 2012 American Geophysical Union.

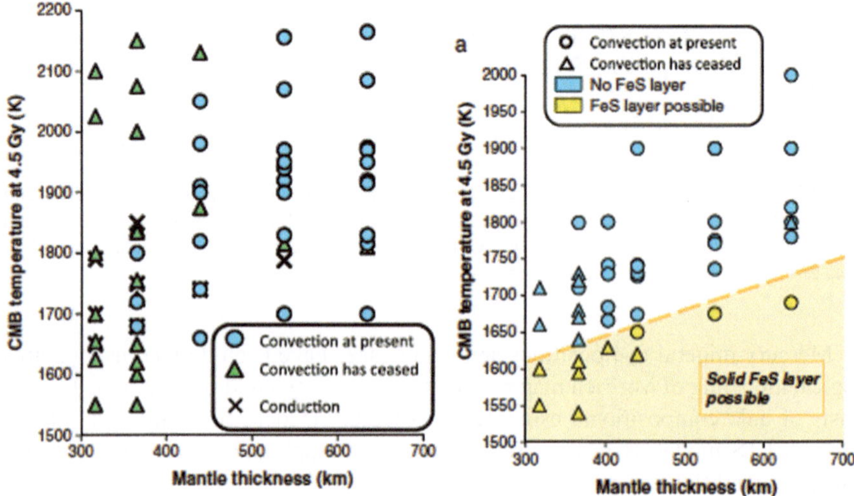

Fig. 2.7 Model for core mantle boundary temperature versus mantle thickness, indicating (left) convection that could persist, providing the mantle is > 400 km. Right: In the presence of a FeS outer shell a mantle of this size could remain convecting until the present. (Michel et al. JGR Planets 2013, Fig. 2 (left) and 6a (right), © 2013 American Geophysical Union)

in the mantle, while U, Th, and K abundances constrain the rate of radiogenic heat reduction. When these data are combined with the determinations of C/MR^2 and C_M/C ratio (Smith et al. 2012), as described below, the implications are that Mercury's core is larger and its mantle is thinner (slightly larger than 400 km) (Hauck et al. 2013) than previously estimated. Michel and coworkers (2013) explored the possibility that convection ceased at some point, and determined that mantle convection could persist as long as the mantle is > 400 km (Fig. 2.7), resulting in widespread high temperature magmas with high degrees of partial melting. The systematic decrease in Mg and sulfides from older to younger terranes is consistent with a decrease in degree of partial melting. The presence of orthopyroxene enstatite, with its high Mg content, as a major component of volcanic deposits, doesn't preclude the presence of clinopyroxene (Stockstill-Cahill et al. 2012). However, the presence of clinopyroxene, which melts at a lower temperature than orthopyroxene and olivine, increases the degree of partial melting to the extent that it is not consistent with the observed decrease in the degree of partial melting.

The most efficient means of heat transport is convection. How well does convection work in Mercury's mantle? A thin mantle implies a low Rayleigh number (ratio between buoyancy and viscous dissipation forces that scale as the third power of the convecting layer) and marginal instability allowing convection. However, higher starting temperature and higher heat production can offset the effect of a thin mantle by lowering its viscosity. Do the large variations in externally induced surface temperature at Mercury's hot poles play a special role? Michel and coworkers (2013) argue that such variations cause larger convection cells with somewhat prolonged melting, and that this does change the thermal structure and thickness of the lithosphere. Long wavelength variations in topography resulting from variations in depth to the core mantle boundary could induce variations in boundary temperature, affecting heat flow from the core and thus the generation and morphology of the magnetic field. An outer FeS solid shell, which would result in a boundary temperature below the melting point of FeS, has been proposed to explain the greater core density than the implied by low iron content (Smith et al. 2012). Such a layer provides even greater constraint than required to explain a mantle that remains convecting over a longer period of time. Current models (Michel et al. 2013) indicate that convection continued until less than 1 billion years ago (Fig. 2.7).

Grott and coworkers (2011) generated thermal models using the best available data on Mercury's bulk composition and a thermally insulating regolith layer. These constraints provide a stiff enough mantle rheology and inefficient enough mantle convection to slow cooling and prolong production of magma until 2.5 billion years after core formation, consistent with MESSENGER observations (Prockter et al. 2010). Volume increase associated with mantle differentiation offsets contraction caused by core cooling to some extent, although an order of magnitude more contraction than observed by *Mariner 10* is required. Crustal thicknesses between 10 to 40 km and high (6 %) core sulfur contents are implied. Grott and coworker (2011) models imply a two-stage magnetic field generation after core formation. Over the first few hundred million years thermal convection occurs in the fluid core, resulting in a thermally driven dynamo. As a small inner core solidifies, the mechanism becomes the composition convection in the non-eutectic fluid outer core, a mechanism consistent with the weak magnetic field observed today (Anderson et al. 2008). Deep dynamo models (Christensen and Wicht 2008) predict such a weak field and inner core but require sulfur content that would result in early core freezing and contraction. On the other hand, recent experiments demonstrate that FeS systems might exhibit melting behavior and a crystallization regime very different from that assumed for the other terrestrial planets (Chen et al. 2008). Iron precipitation may depend on sulfur content, and occur not just at the inner/outer core boundary but elsewhere (double snow state). Such precipitation could be a source of compositional buoyancy that could act as the 'compositional convection' driver for the weak dynamo. The observed crustal thickness is compatible with this model, which allows modest mantle convection that could continue to the present (King 2008; Redmond and King 2007). However, strictly speaking, the chondritic composition source material proposed for Mercury provides too little U and Th to be consistent with this model.

2.5 Surface Features and Implications for the Interior

Though the photogeological and geomorphological details of surface features will be discussed in more detail in Chap. 3, this section discusses surface structural features from the standpoint of deep interior processes.

According to Achille and coworkers (2012) MESSENGER imaging with >50 degree solar incidence angle revealed many undetected surface structures, implying greater spatial distribution and density of tectonic features and thus evidence for greater extent of compressional strain and accompanying radius decrease (2.4–3.6 km) 3–4 times higher than previous estimates.

Lu and coworkers (2011) studied the chaotic terrain area antipodal to the Caloris impact basin to provide information on Mercury's interior structure through observed effects that would have resulted from seismic focusing of waves resulting from the impact. Compression and rarefaction occur as waves propagate through the planet and crustal layer, and disruption occurs when tensile stress exceeds tensile strength. Impacts produced by both surface acoustic (Rayleigh) waves and mantle guided waves trapped between the core and the free surface caused the terrain formation. The terrain consists of 5–10-km-wide hills and depressions 0.1–0.8 km in height. The models predicted disruption zone of 5 angular degrees, but the actual disruption is more like 10 angular degrees, potentially indicating that antipodal response to the impact was modulated or magnified by the shallow outer layer of Mercury.

Caloris, the largest known impact basin on Mercury, has had a broad influence not just on Mercury's surface but potentially on the interior, by affecting heat flow within the mantle and core (Roberts and Barnouin 2013). Perhaps the thermal impulse of its formation changed the underlying dynamics of the thin mantle and was responsible for the region's subsequent volcanism. On the other hand, Roberts and Barnouin (2013) indicate that the subsequent heating could not have penetrated to the core and thus had no impact on the dynamo. The younger age of the volcanic plains around Caloris indicate that they are not impact melt. Color variations among the plains indicate a composition that evolved. Both of these conditions support a volcanic origin for the plains.

Roberts and Barnouin (2013) modeled shock heating and convection in the mantle with standard impact scalings as follows. Similar volumes of melt are produced for a given impact energy. Projectile size determines the extent of heating, and thermal effects on the deep mantle may last for some time. Mantle composition affects density, which also affects melt behavior. Melting occurs where the temperature is hottest beneath a thick stagnant lid, and because the mantle is shallow and gravity and pressure relatively low, melting can occur at significant depth in the mantle. Prior to an impact event, a significant fraction of the lower mantle may already be melted and mixed due to ambient conditions. The impact causes a sharp rise in temperature and subsequent melting due to upwelling material undergoing decompressive melting. Impact heating is confined to the small convective cell closest to the impact site for this thin mantle, heat decaying strongly with distance, confining it to regional rather than global influence.

Impact-induced melting results in a pocket of highly depleted material below, which rises along with the heated region and spreads out along the stagnant lid (Roberts and Barnouin 2013). Thermal buoyancy exceeds chemical buoyancy, resulting in well-mixed lower mantle overlain with the strongly depleted, high viscosity lid, covered with volcanic plains around the impact site. Surrounding this area is well-mixed mantle overlain with unprocessed, undepleted stagnant lid. Pockets of depleted material have been well mixed into the mantle. There is a strong compositional contrast between basement rocks beneath volcanic plains interior to the basin and those surrounding the basin. Although most of the heat dissipates relatively quickly, the stagnant lid cause some heat to be retained for tens of millions of years. A lower mantle viscosity has a modest influence, resulting in a somewhat thinner stagnant lid, a second phase of melting at similar depth, somewhat less contrast between melt compositions and more upwellings, and a somewhat faster cooling mantle. According to Roberts and Barnouin (2013), the key influence is temperature, related to heat externally generated by impact and internally generated through radiogenic heating and not composition.

Exterior melt production is likely only for the slower, larger impacts, and the average impact velocity (42.5 km/s) is far too high for melt production (LeFeuvre and Wieczorek 2008), which come from a much deeper source than conventional impact melts and regions with different compositions, especially in terms of the highly variable Mg. Impact-induced volcanism should be systematically supressed over time as more easily melted components are depleted.

2.6 Figure and Dynamics of Mercury

A number of workers have improved our understanding of the figure and dynamics of Mercury from MESSENGER observations or ground-based observations taken in preparation for the mission (Table 2.3). Margot and coworkers (2012) derived better values for the spin properties, including the position of the spin axis (2.04 ± 0.08 arc minutes with respect to orbit normal) from Earth-based radar observations, indicating Mercury is in or near a Cassini state. An 88-day libration pattern in the rotation rate is due to solar gravity torques acting on the asymmetrically shaped planet. Forced libration amplitude (38.5 ± 1 arc sec) corresponds to 450 m of displacement at the equator. By combining these data with second-degree gravity harmonics derived from MESSENGER measurements by Smith and coworkers (2012), Margot and coworkers (2012) determined the polar moment of inertia (C/MR^2) to be 0.346 ± 0.14 (where C is the moment of inertia, M as the mass, and R is the radius of Mercury), and the fraction of that moment of inertia corresponding to the outer librating shell (C_M/C), from which the size of the core can be estimated, is thought to be 0.431 ± 0.025 (where C_M is the moment of inertia of the outer rigid shell only). This model assumes that the core mantle boundary and inner core boundary are axially symmetric, although convective or tidal processes undoubtedly cause some non-hydrostatic variations in mantle density that, as Gao and Stevenson (2012) suggest, could affect inferred mantle densities by 10–20%.

A number of workers have provided new measurements for Mercury's radius and interpretations for variation in its radius. From MESSENGER measurements, Oberst and coworkers (2011) estimate the radius as 2439 ± 0.69 km, and a nearly identical radius for equatorial and polar radii, suggesting negligible gravitational oblateness and topography rather than gravitational oblateness as the source of uncertainty. Perry and coworkers (2011), using MESSENGER radio occultation, estimated the radius as 2439.2 ± 0.5 km. An offset of Mercury's equatorial plane from the center of figure 600 m has been suggested (Anderson et al. 1996; Smith et al. 2010). RMS roughness obtained from these limb observations (0.8) is comparable to that obtained from laser altimetry (0.9) (Oberst et al. 2011). The large regional depression observed in the stereo terrain model may be correlated with a 1500 km impact basin (Preusker et al. 2011).

2.7 Gravity Anomalies

Smith and coworkers (2010, 2012) derived Mercury's northern hemisphere gravity field (Table 2.3, Fig. 2.8), but because of the high northern latitude periapsis resulting in poor coverage in the southern hemisphere, only a long wavelength field in the southern hemisphere was derived. The geoid has a dynamic range of 200 m. Several large gravity anomalies including mascons (mass concentrations similar to those found for young basins on the Moon) can be seen.

The crust is thicker at lower latitudes, and thinner at higher latitudes and beneath some impact basins (Smith et al. 2012). Northern lowlands are apparently 2 km lower than surrounding terrain, and a broad gravity low is centered over the north pole. A mid-latitude ENE/WSW trending discontinuous upland associated with a weak positive gravity anomaly extends over half of Mercury's circumference. Regional anomalies include a locally elevated region at 68 degrees N, 33 degrees E, Caloris Basin, and the southeast rim of the Sobkou Basin. The positive free air anomaly (gravity corrected for variations in elevation relative to perfect geoid) near Sobkou is associated with the adjacent topographic rise (Fig. 2.8). The Bouguer anomaly (gravity corrected for variations in elevation and mass relative to perfect geoid) shows strong positive anomalies over Sobkou and Budh as well as Caloris,

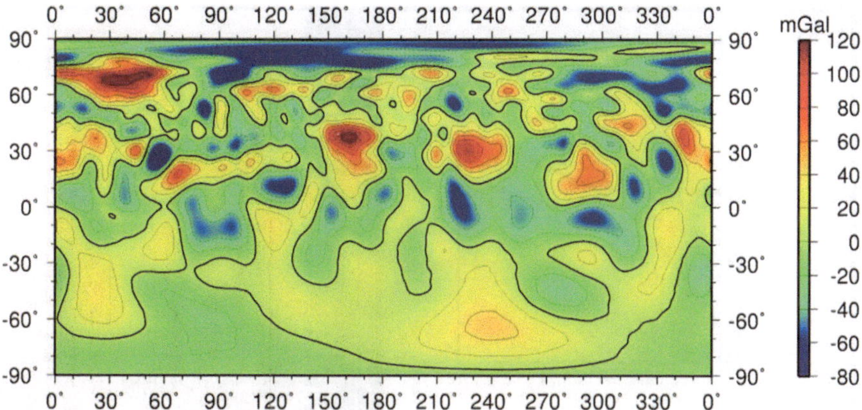

Fig. 2.8 Gravity free air anomalies (compensated for elevation) to harmonic order and degree 50. Note high sensitivity in northern latitudes due to lower altitude observations. (Courtesy of Mazarico et al. 2014, LPS, Fig. 3)

and a weaker negative anomaly over the rise, indicating crustal thinning beneath basins, a phenomenon associated with mascons and indicating that the crust mantle boundary is elevated beneath these basins.

As discussed above, from derivations of planetary spin parameters combined with low degree gravity field a radial density distribution model was created consistent with a solid silicate crust and mantle overlying a solid iron sulfide layer surrounding an iron-rich liquid outer core and solid inner core. Based on the bulk composition implied by the surface geochemical data, the density contrast between crust and mantle is assumed to be approximately 200 kg/m^3. The gravity model (Smith et al. 2012) points to an interior structure for Mercury different from those of other inner Solar System bodies, and when combined with thermal and magnetic field models suggests a solid lower density layer capping the core. Combined bulk compositional data suggest FeS for the solid 'cap' and a core with Fe alloyed with Si as well as S. A static electrically conducting layer surrounding the core decreases overall field strength and attenuates its harmonic components in a manner consistent with observations.

2.8 Mercury's Core

Ground-based radar, *Mariner 10,* and MESENGER observations all confirm, in various ways described here, a partially molten core. The <1 value of the C_M/C (Smith et al. 2012) implies a liquid layer at a depth that decouples the motion of the outer shell from the interior on a time scale of 88 days, a Mercury year. Table 2.3 lists values derived for the dynamic figure properties, including oblateness, the moments of inertia values, and ellipticity, derived from MESSENGER. These agree within one standard deviation with ground-based radar derivations.

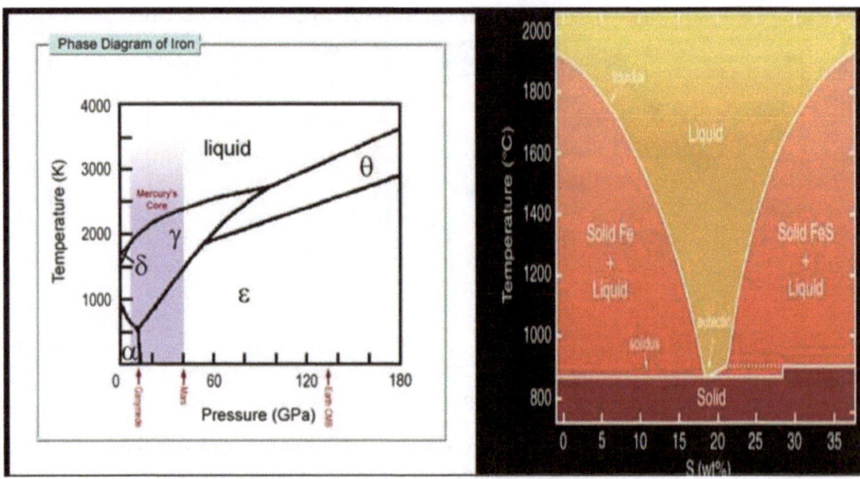

Fig. 2.9 Simplified iron phase diagram (modified from Ahrens et al, 2002) and Fe and S system eutectic diagram. Courtesy of Diviner Team Website Mercury Poster.

Both Mercury and Earth have partially molten cores and dynamos, but with a notable difference in size between the mantles (much smaller on Mercury) and the core (much larger on Mercury), and the very different structure of Mercury's core with its outer molten layer and outer solid 'cap.' Gravity conditions correspond roughly to Mars based on core size, but Ganymede based on planet size. Hauck and coworkers (2013) described the structure of the core and internal structure in detail. The top of the liquid core is at 2020 km, with density decreasing by a factor of 2 immediately below (6980+/–280 kg/m^3) to immediately above (3380+/–200 kg/m^3) this boundary. The discontinuity is persistent for a wide range of compositions. Mercury has a solid outer shell, with a thickness of 410+/–36 km and an average density of 3650+/–225 kg/m^3. Observed compositions favor partitioning of Si and/or S into the metallic core during global geochemical differentiation (Hauck et al. 2013). The thermodynamic properties of a core with both Si and S suggest formation of an FeS-rich layer that is at least partially solid during the core formation process, as illustrated by the iron phase diagram (Ahrens et al. 2002), eutectic of the Fe and S system (Fig. 2.9). Increasing S abundance in the core translates into a smaller solid inner core increasing in size as a function of time (Fig. 2.10). Based on the current inner core size, a larger S abundance seems more likely.

These combined data led Hauck and coworkers (2013) to interpret the most likely core composition to be Fe-S-Si, although moments of inertia measurements are compatible with FeS, FeSi or Fe-S-Si systems (Fig. 2.11). FeSi, Fe-S-Si, and FeS vary in oxygen fugacity and reduction from most to least, respectively. FeS immiscibility and buoyancy drives the chemical segregation of a solid FeS 'cap' on the core and allows that core to remain at least partially molten to the present day, increasing the density of the outer shell. Some S partitions to silicate melt during differentiation. The immiscibility of sulfides in the core leads to the concentrated

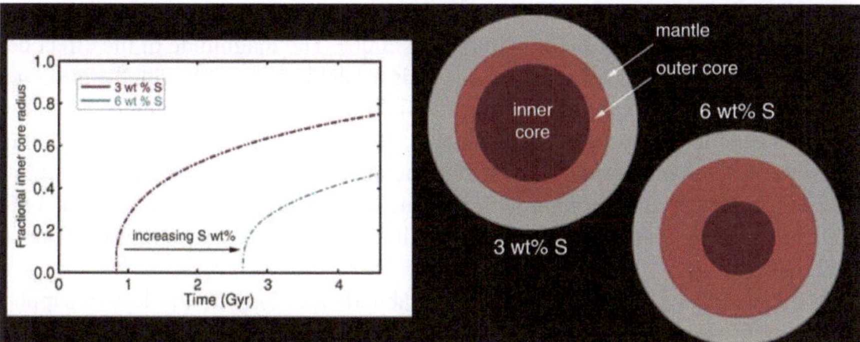

Fig. 2.10 S abundance in the core with implications for core size as a function of time. (Courtesy of Diviner Website Mercury Poster)

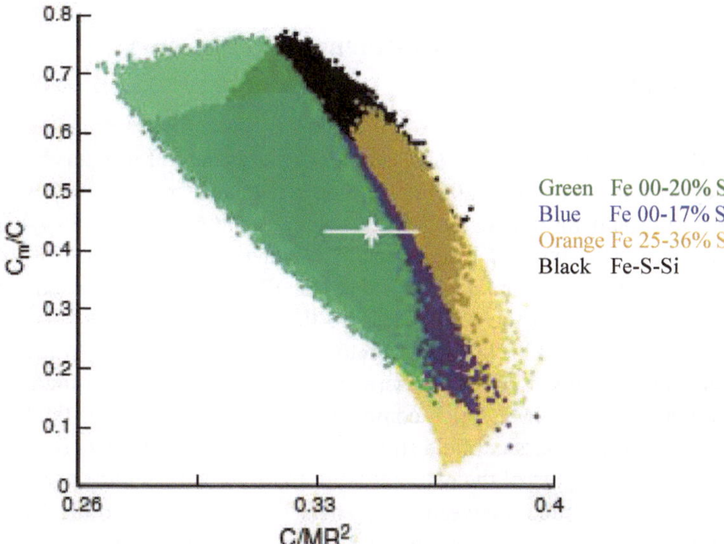

Fig. 2.11 Core composition as a function of C/MR2 and CM/C, with key indicated on diagram. (Hauck et al. 2013, Fig. 2, © 2013 American Geophysical Union)

FeS layer on top, favoring precipitation of FeS-rich solids on top and more Si rich materials below. As the radius of the solid/liquid boundary increases, C/MR^2 increases, and C_M/C decreases (Fig. 2.11). Increase in the solid outer shell density leads to increases in both, as it accounts for a larger proportion of the planet's mass. According to this work, earlier estimates of crust and mantle thickness were too high.

Composition stratification in the Fe-S-Si core affects magnetic field generation as well. The existence of the stable, electrically conductive layer beneath the core mantle boundary attenuates the shorter term, and shorter wavelength components of the dynamo-generated field (Christensen 2006). The magnitude of the effect depends on the thickness and conductivity. This model is consistent with the observed long-wavelength structure of the magnetic field with apparent attenuation of higher than quadrupole harmonics (Anderson et al. 2011). The inner core must be small to allow the convecting region in the core to be of sufficient thickness. Large solubility of Si in Fe metal combined with the smaller melting point depression of FeSi systems, indicate that a core with Si could be larger than one with S alone (Hauck et al. 2004).

MESSENGER GRS-derived U and Th abundances may be too low to support strong partitioning into the core and cause additional heating. High S abundance would favor K solubility and its distribution as a radiogenic element to increase core heating and aid in dynamo generation, as well as slowing cooling and delaying the onset of the solid FeS 'cap.'

2.9 Internal and External Influences on Mercury's Dynamo

Conventional dynamo theory doesn't explain the weakness of Mercury's magnetic field. Heyner and coworkers (2011) proposed a feedback mechanism between the magnetic field and the magnetosphere to account for Mercury's observed weak magnetic field. The external Interplanetary Magnetic Field (IMF) contributes to both the magnetosphere and interior magnetic fields. Mercury's internally generated field lies close to the surface due to the intensity of the solar wind flux. Heyner and coworkers (2011) use magneto-hydrodynamic models to invoke coupling between Mercury's magnetosphere and internal dynamo, and found that the combination of typical IMF fields combined with Mercury's weak field matches the observations. The classical field strength saturation mechanism (Lorentz Force) is replaced by the impact of the external field. In effect, a magnetopause is created as a result of interactions between the planetary dipole field and the solar wind. According to Heyner and coworkers (2011), the external field resulting from the magnetopause current reaches into the fluid outer core, where the dynamo dipole field and external field are anti-parallel, generating anticyclones that convert the poloidal field into an azimuthal field, stretching it away from the equator, mixing and weakening both fields, and causing a less pronounced strengthening at the north pole and more uneven distribution. Heyner and coworkers (2011) also suggest that inner-core crystallization even after the planet's heat flux became sub-adiabatic, which may have revitalized Mercury's dynamo. Anderson and coworkers (2011) attempted to separate internal and external magnetic field contributions, and found that the global field is southward directed and spin aligned, with the dipole offset along the spin axis by 484+/–11 km north of the geographic equator, as discussed in Chap. 4. As a result, the magnetic field at the surface is 3–4 times larger in the northern hemisphere. The surface area open to direct magnetic flux is 4 times larger in the southern hemisphere, which is likely to result in greater sputtering at the surface and preferential precipitation of plasmas. Saturn is similarly aligned and is offset along its dipole, but ratio of offset to diameter is 5 times greater for Mercury. Saturn's offset is thought to result from a differentially rotating conducting layer between a deeper tilted field and exterior fields, perhaps indicating an analogous mechanism for Mercury.

References

Achille, G., Popa, C., Massironi, M., Mazzotta Epifani, E., Zusi, M., Cremonese, G., Palumbo, P.: Mercury's radius change estimates revisited using MESSENGER data. Icarus. **221**, 450–460 (2012)

Ahrens, T., Holland, K., Chen, G.: Phase diagram of iron, revised-core temperatures. Geophys. Res. Lett. **29**(7), 54-1–54-4 (2002)

Anderson, B., Acuna, M., Korth, H., Purucker, M., Johnson, C., Slavin, J., Solomon, S., McNutt, R.: The structure of Mercury's magnetic field from MESSENGER's first flyby. Sci. **321**, 82–85 (2008)

Anderson, B., Johnson, C., Korth, H., Purucker, M., Winslow, R., Slavin, J., Solomon, S., McNutt, R., Raines, J., Zurbuchen, T.: The Global magnetic field of mercury from MESSENGER orbital observations. Sci. **333**, 1859–1862 (2011)

Anderson, J.D., Jurgens, J., Lau, E., Slade, M., Schubert, G.: Shape and orientation of Mercury from radar ranging data. Icarus. **124**, 690–697 (1996)

Brown, S., Elkins-Tanton, L.: Compositions of Mercury's earliest crust from magma ocean models. Earth. Planet. Sci. Lett. **286**, 446–455 (2009)

Chen, B., Li, J., Hauck, S.: Non-ideal liquidus curve in the Fe-S system and Mercury's snowing core. Geophys. Res. Lett. **35**, L07201 (2008). doi:10.1029/2008GL033311

Christensen, U.: A deep dynamo generating Mercury's magnetic field. Nat. **105**, 1058 (2006). doi:10.1038/nature05342

Christensen, U., Wicht, J.: Models of magnetic field generation in partly stable planetary cores: applications to Mercury and Saturn. Icarus. **196**, 16–34 (2008)

Clark, P.: Mercury's interior, in dynamic planet: Mercury in the context of its environment, pp. 37–60, Springer, Berlin (2007).

Denevi, B., Robinson, M, Solomon, S., Murchie, S., Blewett, D., Domingue, D., McCoy, T, Ernst, C., Head, J., Watters, T., Chabot, N.: The evolution of Mercury's crust: a global perspective from MESSENGER. Science. **324**, 613–618 (2009)

Diviner Team: Mercury Website. http://diviner.ucla.edu/mercury/posters/Poster-13/poster-13.pdf (2012). Accessed 13 Nov 2014

Ebel, D., Alexander, C.M.: Equilibrium condensation from Chondritic porous IDP enriched vapor: Implications for Mercury and enstatite chondrites origins. Plan. Space. Sci. **59**, 1888–1894 (2011)

Evans, L., Peplowski, P., Rhodes, E., Lawrence, D., McCoy, T., Nittler, L., Solomon, S., Sprague, A., Stockstill-Cahill, K., Starr, R., Weider, S., Boynton, W., Hamara, D., Goldsten, J.: Major element abundances on the surface of Mercury: Results from the MESSENGER Gamma-ray Spectrometer. J. Geophys. Res. Planet. **117**, E00L07 (2012). doi:10.1029/2012JE004178

Gao, P., Stevenson, D.: The effect of nonhydrostatic features on the interpretation of Mercury's mantle density from MESSENGER results. American Astronomical Society DPS meeting 401.08, DPS 44 (2012)

Griffin, A.A., Millman, P.M., Halliday, I.: "The fall of the Abee meteorite and its probable orbit". R. Astron. Soc. Can. J. (ISSN 0035-872X), **86**(8), 5–14 (1992)

Grott, M., Breuer, D., Laneuville, M.: Thermo-chemical evolution and global contraction of Mercury. Earth. Planet. Sci. Lett. **307**, 135–146 (2011)

Hauck, S., Dombard, A., Phillips, R., Solomon, S.: Internal and tectonic evolution of Mercury. EPSL. **222**, 713–728 (2004)

Hauck, S., Margot, J., Solomon, S., Phillips, R., Johnson, C., Lemoine, F., Mazarico, E., McCoy, T., Padovan, S., Pele, S., Perry, M., Smith, D., Zuber, M.: The curious case of Mercury's internal structure. JGR Planets. **118**, 1204–1220 (2013)

Head, J., Chapman, C., Strom, R., Fassett, C., Denevi, B., Blewett, D., Ernst, C., Watters, T., Solomon, S., Murchie, S., Prockter, L., Chabot, N., Gillis-Davis, J., Whitten, J., Goudge, T., Baker, D., Hurwitz, D., Ostrach, L., Xiao, Z., Merline, W., Kerber, L., Dickson, J., Oberst, J., Byrne, P., Klimczak, C., Nittler, L.: Flood Volcanism in the Northern High Latitudes of Mercury revealed by MESSENGER. Science. **333**, 1853–1856 (2011)

Heyner, D., Wicht, J., Gomez-Perez, N., Schmitt, D., Auster, H.-U., Glassmeier, K.-H.: Evidence from numerical experiments for a feedback dynamo generating mercury's magnetic field. Sci. **334**, 1690–1693 (2011)

Keil, K.: Mineralogical and chemical relationships among enstatite chondrites. J. Geophys. Res. Planet. **73**, 6945–6976 (1968)

King, S.: Pattern of lobate scarps on Mercury's surface reproduced by a model of mantle convection. Nat. Geosci. **1**, 229–232 (2008). doi:10.1038/ngeo152

Kerber, L, Head, J., Solomon, S., Murchie, S., Blewett, D., Wilson, L.: Explosive volcanic eruptions on Mercury: eruption conditions, magma volatile content, and implications for interior volatiles abundances. EPS., **285**, 263–271 (2009)

Kerber, L., Head, J., Blewett, D., Solomon, S., Wilson, L., Murchie, S., Robinson, M., Denevi, B., Domingue, D.: The global distribution of pyroclastic deposits on Mercury: The view from MESSENGER flybys 1–3. Planet. Sp. Sci. **59**, 185–1909 (2011)

LeFeuvre, M., Wieczorek, M.: Nonuniform cratering of the terrestrial planets. Icarus. **197**, 291–306 (2008)

Lu, J., Sun, Y., Toksoz, M.N., Zheng, Y., Zuber, M.: Seismic effects of the Caloris basin impact, Mercury. Plan. Space. Sci. **59**, 1981–1991 (2011)

Margot, J., Peale, S., Solomon, S., Hauck, S., Ghigo, F., Jurgens, R., Yseboodt, M., Giorgini, J., Padovan, S., Campbell, D.: Mercury's moment of inertia from spin and gravity data. J. Geophys. Res. Planet. **117**, E12 (2012). doi:10.1029/2012JE004161

Mazarico, E., Genova, A., Goossens, S., Lemoine, F., Smith, D., Zuber, M., Neumann, G., Solomon, S.: The gravity field of Mercury from MESSENGER, Lunar and Planetory Science Conference, 45, 1863.pdf (2014).

McCoy, T., Dickinson, T., Lofgren, G.: Partial melting of the Indarch (EH4) meteorite: A textural, chemical, and phase relations view of melting and melt migration. Meteorit. Planet. Sci. **34**, 735–746 (1999)

McCubbin, F., Riner, M., VanderKaaden, K., Burkemper, L.: Is Mercury a volatile rich planet? Geophys. Res. Planet. **39**, L09202 (2012). doi:10.1029/2012GL051711

Michel, N., Hauck, S., Solomon, S., Phillips, R., Roberts, J., Zuber, M.: Thermal evolution of Mercury as constrained by MESSENGER observations. J. Geophys. Res. Planet. **118**, 1033–1044 (2013)

Nittler, L., Starr, R., Weider, S., McCoy, T., Boynton, W., Ebel, D., Ernst, C., Evans, L., Goldsten, J., Hamara, D., Lawrence, D., McNutt, R., Schlemm, C., Solomon, S., Sprague, A.: The major element composition of Mercury's surface form MESSENGER X-ray spectrometry. Sci. **333**, 1847–1849 (2011)

Oberst, J., Elgner, S., Turner, F.S., Perry, M., Gaskell, R., Zuber, M., Robinson, M., Solomon, S.: Radius and limb topography of Mercury obtained form images acquired during the MESSENGER flybys. Plan. Space. Sci. **59**, 1918–1924 (2011)

Peplowski, P., Evans, L., Hauck, S., McCoy, T., Boynton, W., Gillis-Davis, J., Ebel, D., Goldsten, J., Hamara, D., Lawrence, D., McNutt, R., Nittler, L., Solomon, S., Rhodes, E., Sprague, A., Starr, R., Stockstill-Cahill, K.: Radioactive elements on Mercury's surface from MESSENGER: Implications for the planet's formation and evolution. Sci. **333**, 1850–1852 (2011)

Peplowski, P., Lawrence, D., Rhodes, E., Sprague, A., McCoy, T., Denevi, B., Evans, L., Head, J., Nittler, L., Solomon, S., Stockstill-Cahill, K., Weider, S.: Variations in the abundances of potassium and thorium on the surface of Mercury: Results from the MESSENGER Gamma-ray spectrometer. J. Geophys. Res. Planet. **117**, E00L04 (2012a). doi:10.1029/2012JE004141

Peplowski, P., Rhodes, E., Hamara, D., Lawrence, D., Evans, L., Nittler, L., Solomon, S.: Aluminum abundance on the surface of Mercury: Application of a new background-reduction technique for the analysis of gamma-ray spectroscopy data. J. Geophys. Res. Planet. **117**, E00L10 (2012b). doi:10.1029/2012JE004181

Perry, M., Kahan, D., Barnouin, O., Ernst, C., Solomon, S., Zuber, M., Smith, D., Phillips, R., Srinivasan, D., Oberst, J., Asmar, S.: Measurement of the radius of Mercury by radio occultation during the MESSENGER flybys. Plan. Space Sci. **59**, 1925–1931 (2011)

Preusker, F., Oberst, J., Head, J., Watters, T., Robinson, M., Zuber, M., Solomon, S.: Stereo topographic models of Mercury after three MESSENGER flybys. Plan. Space Sci. **59**, 1920 − 1917 (2011)

Prockter, L., Ernst, C., Denevi, B., Chapman, C., Head, J., Fassett, C., Merline, W., Solomon, S., Watters, T., Strom, R., Cremonese, G., Marchi, S., Massironi, M.: Evidence for young volcanism, on Mercury from the thirds MESSENGER flyby. Science. **329**(5992), 668–671 (2010)

Purucker, M., Sabaka, T., Solomon, S., Anderson, B., Korth, H., Zuber, M., Neumann, G.: Mercury's internal magnetic field: constraints on large and small-scale fields of crustal origin. Earth. Planet. Sci. Lett. **285**, 340–346 (2009)

Redmond, H., King, S.: Does mantle convection currently exist on Mercury? Phys. Earth. Plan. Inter. **164**(3–4), 221–231 (2007)

Rhodes, L., Evans, L., Nittler, L., Starr, R., Sprague, A., Lawrence, D., McCoy, T., Stockstill-Cahill, K., Goldsten, J., Lawrence, D., McCoy, T., Stockstill-Cahill, K., Goldsten, J., Peplowski, P., Hamara, D., Boynton, W., Solomon, S.: Analysis of MESSENGER gamma-ray spectrometer data from the Mercury flybys. Plan. Space. Sci. **59**, 1829–1841 (2011)

Riner, M., McCubbin, F., Lucey, P., Taylor, G.J., Gillis-Davis, J.: Mercury surface composition: Integrating petrologic modeling and remote sensing data to place constraints on FeO abundance. Icarus. **209**, 301–313 (2010)

Roberts, J., Barnouin, O.: The effect of Caloris impact on the mantle dynamics and volcanism of Mercury. J. Geophys. Res. Planet. **117**, E02007 (2013). doi:1029/2011JE003876

Stockstill-Cahill, K., McCoy, T., Nittler, L., Weider, S., Hauck, S.: Magnesium-rich crustal compositions on Mercury: Implications for magmatism for petrologic modeling. J. Geophys. Res. Planet. **117**, E00L15 (2012). doi:10.1029/2012JE004140

Smith, D., Zuber, M., Phillips, R., Solomon, S., Neumann, G., Lemoine, F., Peale, S., Margot, J., Torrence, M., Talpe, M., Head, J., Hauck, S., Johnson, C., Perry, M., Barnouin, O., McNutt, R., Oberst, J.: The equatorial shape and gravity field of mercury from MESSENGER flybys 1 and 2. Icarus. **209**, 88–100 (2010)

Smith, D., Zuber, M., Phillips, R., Solomon, S., Hauck, S., Lemoine, F., Mazarico, E., Neumann, G., Peale, S., Margot, J.-L., Johnson, C., Torrence, M., Perry, M., Rowlands, D., Goossens, S., Head, J., Taylor, A.: Gravity field and internal structure of Mercury from MESSENGER. Sci. **336**, 214–217 (2012)

Sprague, A., Hunten, D., Lodders, K.: Sulfur at Mercury, elemental at the poles and sulfides in the regolith. Icarus. **118**, 211–215 (1995)

Sprague, A., Emery, A., Donaldson, K., Russell, R., Lynch, D., Mazuk, A.: Mercury: Mid-Infrared (3-13 u) observations show heterogeneous composition, presence of intermediate and basic soil types, and pyroxene. Meteorit. Planet. Sci. **37**, 1255–1268 (2002)

Sprague, A., Massey, S.: Mercury's exosphere: A possible source of Na. Planet. Space. Sci. **5**(11), 1614–1621 (2007)

Stevenson, D., Lunine, J.: Rapid formation of Jupiter by diffusive redistribution of water vapor in the solar nebular. Icarus. **75**, 146–155 (1988)

Weider, S., Nittler, L., Starr, R., McCoy, T., Stockstill-Cahill, K., Byrne, P., Denevi, B., Head, J., Solomon, S.: Chemical heterogeneity of Mercury's surface revealed by the MESSENGER X-ray spectrometer. J. Geophys. Res. Planet. **117**, E00L05 (2012). doi:10.1029/2012JE004153

Weider, S., Nittler, L., Starr, R., McCoy, T., Solomon, S.: Variations in abundance of iron on Mercury's surface from MESSENGER X-ray spectrometer observations. Icarus. **235**, 170–186 (2014)

Zolotov, M., Sprague, A., Hauck, S., Nittler, L., Solomon, S., Weider, S.: The redox state, FeO content, and origin of sulfur-rich magmas on Mercury. J. Geophys. Res. Planet. **118**, 138–146 (2013)

Chapter 3
Mercury's Surface

3.1 Details of Mercury's Surface Features and Their Implications

The MESSENGER mission has provided far more detailed coverage of Mercury's surface, particularly its northern hemisphere, than previously available, including photomosaic, spectral reflectance, surface composition, and topography maps. The result is a far better basis for understanding formation histories and mechanisms of its bombardment, tectonic, and volcanic features and terrains, plus volatile deposits. Detailed discussion of elemental abundance measurements, from which bulk abundances and thus interior composition and origin can be inferred, can be found in Chap. 2.

3.2 Surface Thermal Environment

Mercury's surface temperature ranges from 700 K at the hot pole to 100 K at the cold pole at perihelion. Figure 3.1 (BepiColombo STR 2000) illustrates Mercury's temperature as a function of latitude, longitude, and its position in orbit. At perihelion, the orbital angular motion exceeds its rotation rate so that the Sun stays above the meridian for a longer time, causing extreme heating at the hot pole and extreme cooling at the cold pole. At aphelion, the converse is true: the rotation rate exceeds the orbital angular motion. This motion also has an impact on solar wind interaction, radiation exposure and damage of the surface, and the magnetosphere, potentially compressing it all the way to the surface at perihelion (Langevin 1997), an effect that will be discussed in more detail in Chap. 4.

Thermal properties of the regolith are estimated in Table 3.1 (BepiColumbo STR 2000; Kring 2006), indicating a lunar-like regolith with lower overall albedo. Yan and coworkers (2006) have speculated that Mercury's high temperatures could lead to 'thermal annealing' that could compensate for 'radiation damage,' the physical disruption of the individual grains.

P. E. Clark, *Mercury's Interior, Surface, and Surrounding Environment*, SpringerBriefs in Astronomy, DOI 10.1007/978-1-4939-2244-4_3

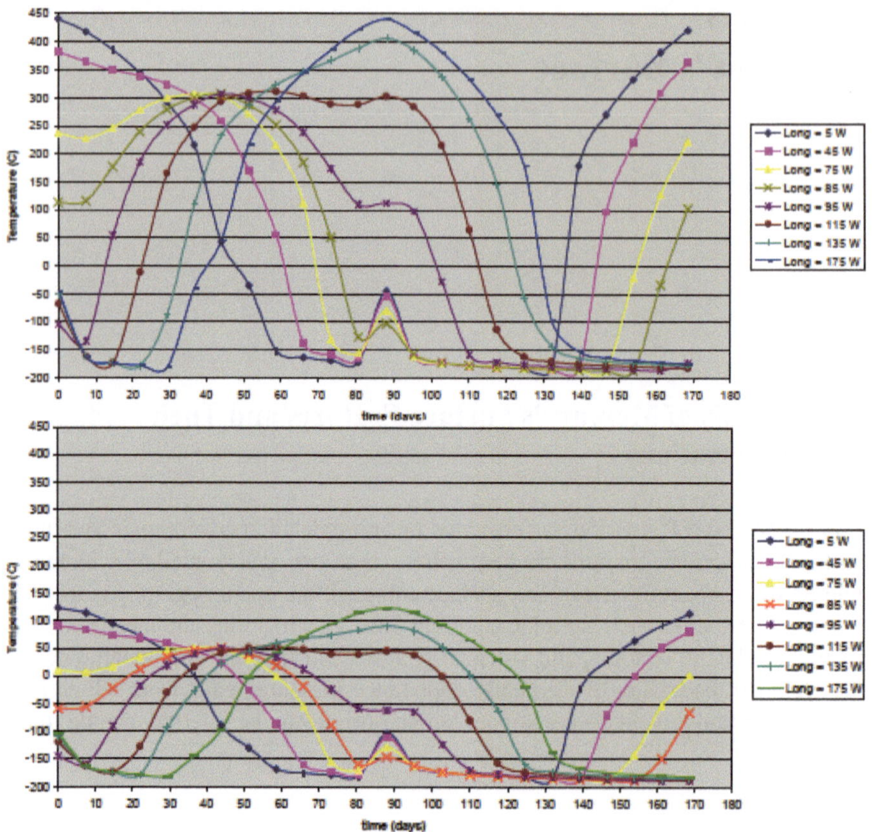

Fig. 3.1 Model temperature as a function of longitude during the course of Mercurian year for 0 (*top*) and 85 (*bottom*) degrees latitude. (Courtesy of ESA, BepiColombo STR 2000)

Table 3.1 Mercury's thermal regolith properties. (Source: Mercury data, BepiColombo STR 2000; lunar data, Kring 2006)

Body	Albedo	Emissivity	Density	Specific heat	Conductivity W/Mk @1 m depth equator
Mercury	0.07	0.9	1300 kg/m^3	800 j/kg-K	3×10^{-4} (370 K)
Moon	0.11	0.9	1350 kg/m^3	760 j/kg-K	1.5×10^{-4} (220 K)

3.3 Surface Color, Spectral, and Scattering Properties

Whereas color difference maps derived from *Mariner 10* images showed regional color differences but no clear association between color and terrain, MESSENGER WAC-derived color composite maps of Mercury show color differences at

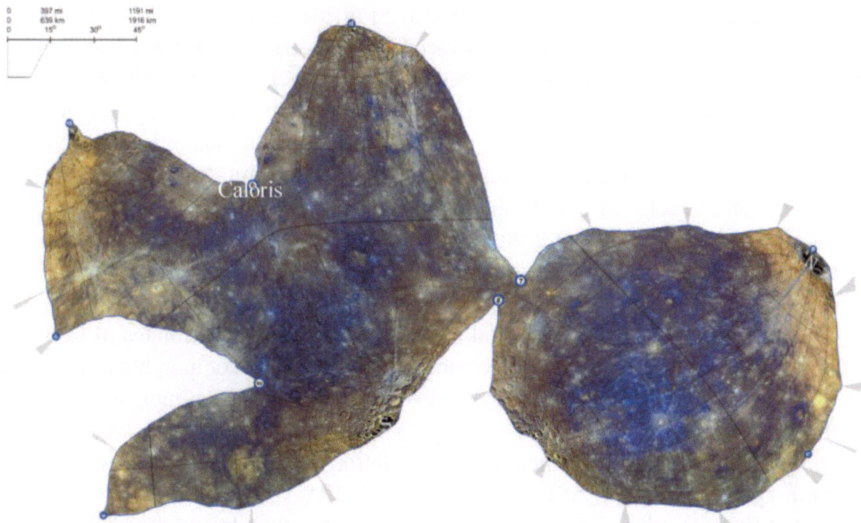

Fig. 3.2 CSNB version of MESSENGER Mercury global photomosaic (Courtesy of NASA/APL), using centroids in orange terrain between occurrences of blue terrain as boundaries (Clark and Clark, 2014). Courtesy of Chuck Clark and Pamela Clark, 2014.

geomorphological boundaries in the MESSENGER global color photomosaic (Fig. 3.2) derived from principal component analysis (first principal component green, second principal component red, 430/1,000 nm ratio blue). Generally, orange denotes higher iron content and blue lower iron content. Darker areas are apparently volcanic, with darker blue indicating older, less abundant iron composition with more opaques, and darker orange indicating younger, more abundant iron composition. Bright blue materials, clearly associated with rays, indicate younger impact material, and bright orange materials are younger volcanic plains with pyroclastics. Brightness is an indicator of fresher, rougher, in the case of impact more comminuted, and thus brighter, components (Kerber et al. 2009).

As Domingue and coworkers noted (2010) evidence for an Fe-bearing mineral absorption feature that has been generally lacking, and the exceptions may indicate heterogeneity. Recent ground-based spectral observations (Erard et al. 2011), which had no clear 1 or 2 μm features, provide an upper limit of 0.15–0.6 % FeO abundance consistent with the presence of enstatite, with implications that spectral differences are caused by different size distributions of regolith particles, including more micron-sized particles. However, the enstatite-bearing asteroids that could be considered analogs of Mercury (Burbine et al. 2002) are distinctively brighter and flatter in the infrared than Mercury spectra. Erard and coworkers (2011) proposed another mineralogical phase to account for darkening and reddening. However, direct compositional measurements indicate significant modifications due to high degree partial melting of enstatite chondrite source material (see Chap. 1). The presence of smaller particles would be supported by high temperature and more intensive flux processes, show darker spectra with minimal reddening, and be consistent with observations.

Domingue and coworkers (2010, 2011a, b) affirm that Mercury's red spectral slope is typical of space-weathered surfaces. Photometric measurements indicate a smoother surface with more compact regolith and higher geometric albedo (reflectance at zero phase angle), but lower overall albedo, than the Moon. Younger units on the Moon are 1.5 times higher in reflectance than analogous units on Mercury, but the difference in albedo between younger and older units is a factor of 1.5 on the Moon, and a factor of 2.0 on Mercury (Denevi and Robinson 2008). Domingue and coworkers (2010) interpret the higher geometric albedo of Mercury as due to shorter reflection paths in the regolith, with scattering centers concentrated near the surface of grains.

Unconstrained by direct elemental abundance data, rough estimates of iron and titanium derived from spectral reflectance observations are comparable to Luna 16 and 24 (averaging 6% and 1%, respectively) and too high (Domingue et al. 2010). Although titanium, with its relatively high neutron cross-section, was proposed in the absence of iron to explain neutron cross-section observed by the MESSENGER neutron spectrometer, MESSENGER XRS and GRS spectrometer measurements indicated titanium abundances too low to account for the observed neutron cross-section (See Chap. 2). The proposal that early forming ilmenite would remove ferrous iron from the magma ocean as an explanation of very low iron silicate abundances is not valid.

Older lunar soils have a more pronounced red slope than older Mercury soils. Mercury has a micrometeoroid flux 5.5 times greater, with a velocity 1.6 times greater, resulting in 13.5 times more melt and 19.5 times more vapor per unit area (Cintala 1992). The higher solar wind particle flux is periodically channeled directly to the surface. Interaction between the IMF and Mercury's magnetic field results in magnetic reconnection, further accelerating solar wind particles along magnetic field lines toward the poles. In addition, surface temperatures are far higher on Mercury. Noble and coworkers (2007) predicted that these conditions could result in more efficient Ostwald ripening and growth of larger grains resulting in a more rapid and efficient production of agglutinates with more scatter centers near the grain surface, and thus in greater darkening and less reddening, as observed (Holsclaw et al. 2010).

Domingue and coworkers (2010, 2011a, b) performed an extensive analysis of the optical properties of Mercury's regolith at visible wavelengths. Scattering centers result from physical boundaries such as inclusions, bubbles, and fractures that change the index of refraction as a function of wavelength and phase angle. A variety of Hapke Model-predicted and observed scattering parameters were used in the analysis (Domingue et al. 2010). Comparison of the abundance of scattering centers (b) and proportion of multiple scatterers (c) for Mercury, the Moon and asteroids (Figure 3.3) are interpreted as follows (Domingue et al. 2011a). The lunar surface exhibits characteristic phase reddening. The contribution of multiple scattering increases as the phase angle increases and as the wavelength increases, and back scattering increases as multiple scattering increases. For Mercury, the back scattering component decreases with increasing wavelength, while the forward-scattering component increases with increasing wavelength, consistent with its observed lack of phase reddening. The general trend is a decrease in scatter at longer wavelengths, indicative

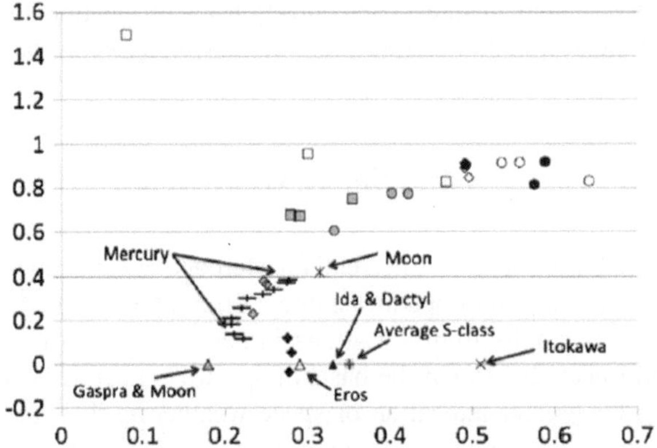

Fig. 3.3 Relationship between scattering parameters and particle structure characteristics for Mercury, named asteroids, the Moon; smooth, rough, and pitted spheres with no internal structure (*open*, *gray*, and *black* circles); low, moderate, and high densities of internal scatterers (*open*, *gray*, and *black* diamonds; irregularly shaped and agglutinate particles (*open* and *gray* squares). (Reprinted from Planetary and space science, p. 59, Domingue et al., 2011a, Figure 8, © 2011, with permission from Elsevier)

of the small size (<400 nm) of scatter centers on the scale of visible wavelengths. The Moon and Mercury have comparable abundances of internal scattering centers, whereas asteroids have a greater abundance of internal scattering centers. Asteroids have a high proportion, the Moon a moderate to high proportion, and Mercury moderate to low abundance of multiple scattering centers (inversely related to c). The size and density of scattering centers are related to the processes that formed them. Thus, these differences indicate structural differences in the Mercury regolith grains that Domingue and coworkers (2010) agree results in part from increased production of space-weathered impact glasses due to the combination of greater flux and velocities of incoming material (Cintala 1992), charged particles, and magnetic flux (Slavin et al. 2010), but also from increased abundance of fractures, cracks, and lattice defects near the surface.

On the meter to millimeter scale, surface roughness varies as a function of wavelength and target. On the meter scale, Mercury is more like smaller asteroids, and smoother than the Moon; on the scale of individual grain structures, including the role of scattering centers (shorter wavelengths), Mercury is more like the Moon (e.g., Hartman and Domingue 1998; Helfenstein and Veverka 1987).

Initial spectral reflectance measurements (e.g., Blewett et al. 2009) identified higher reflectance plains (HRP), intermediate reflectance terrain (IT), and low reflectance material (LRM). They also identified somewhat brighter and redder spots associated with pyroclastics, and brighter bluer crater floor material 'not associated with immature material.' The small and unimodal nature of the variations in color are thought to result from small variations in already low abundance ferrous iron and small variations associated with age. Variations in opaque phases are thought to have the major impact on spectral character. Are opaque phases the source of iron in space-weathering processes? Elemental abundance data discussed in Chap. 2

eliminate the conventional opaques based on very low iron and titanium abundances, but indicate that the iron in iron sulfide could be playing a role in space weathering.

3.4 Regolith Processes and Space Weathering

Indeed, space weathering style is 'different' on Mercury, and albedo variations can't be explained by the addition of opaques alone. Differential accumulation of space-weathering products results from variations in surface composition (Riner et al. 2011; Riner and Lucey 2012). On the basis of spectral signature, younger soils on Mercury have a far greater indication of space weathering than younger soils on the Moon, indicating a far more rapid rate of weathering consistent with Mercury's higher surface temperature and far more frequent encounters with higher speed particles. Surface materials range from brighter with steeper slope (red) to darker with shallower slope (blue), whereas on the Moon, reddening accompanied by darkening is a feature of space weathering. The presence of larger iron metal particles associated with agglutinates alone (Britt Pieters—BP—particles) darkens without reddening. Such a spectral signature is observed in the laboratory as the ratio of BP particles to nanophase iron increases, and would imply that the BP particle to nanophase iron ratio is higher on Mercury than the Moon. According to Riner and coworkers (2011), more mature LRM, IT, and HRP have progressively lower iron-derived maturity indeces, implying variations in initial iron abundance consistent with elemental abundance measurements discussed in Chap. 2.

Brightness contrasts on Mercury are apparently due in large part to differences in surface aging processes known as 'maturity,' as the planet lacks the inherent albedo contrasts between terrains found on the Moon. These regolith processes include comminution (fragmentation) by direct impact, causing bright, young regolith deposition on the surface. Cumulative impacts also result in 'gardening,' or overturn of the regolith. Further erosion of grain surfaces by micrometeorite bombardment causes microcrater formation. This process competes with agglutination (aggregation of grains through partial melting) through partial melting. Glassy microspherules are formed through pyroclastic or impact inducted explosive processes locally. Vapor or liquid phase deposition of nanophase iron may occur on grain surfaces, a process conventionally known as space weathering. The more intense space weathering process should lead to the production of nanophase iron even in relatively low iron soil. The rate of energy deposition from small impacts is 10 times higher for Mercury than the Moon. This is equivalent to a ten times greater increase in exposure time and maturity index, a smaller mean grain size, and a larger proportion of glassy particles (Langevin 1997).

Surface darkening on Mercury may occur, at least in part, by means other than the conventional 'space weathering' models derived for lunar materials (Kerber et al. 2009, 2011). Darkening from these has been postulated to involve reaction of solar wind particles with iron on the Moon. What is the mechanism of darkening on Mercury, in the absence of ferrous iron? Proposed mechanisms include (1) a more aggressive space weathering environment due to ten times higher solar flux at Mercury, requiring much less iron to operate, and (2) elemental carbon (from enstatite chondrite) reduction of other oxide components of the silicate melt to create native metals (e.g., Ni, Co, Si) and generate pyroclastic process drivers CO and CO_2. This is not as unlikely as it may seem. On the Moon, pyroclastic eruptions are driven by CO created by reaction between graphite and FeO in a reducing environment, and metallic Fe is a by-product (e.g., Rutherford and Papale 2009). In order to partition into the gas phase, sulfur requires the presence of H_2O, CO, or CO_2 (Nicholis and Rutherford 2005). In any case, Mercury requires more than twice the volatile abundance of the Moon to drive pyroclastic emplacement to equivalent distances, indicating a much larger volatile inventory consistent with bulk abundances derived from MESSENGER orbital measurements of composition (Kerber et al. 2011). The apparent global distribution of pyroclastic deposits on Mercury may indicate relative uniformity in crustal thickness and/or in internal heat-producing elements, and the lack of smaller deposits may be a resolution effect, or a result of more widespread lower viscosity volcanic flows (Kerber et al. 2011).

Implications from the observations of asteroid Itokawa samples are that nanometer-scale surface deposits may not be metallic iron but iron/magnesium sulfides (Domingue et al. 2014). Unexpected volatile retention may result from more extensive radiation damage, creating 'gas reservoirs' and adsorption surfaces in the finer grained (<45 μm) fraction, which is proportionally larger in Mercury's regolith.

3.5 Surface Topography

MESSENGER altimetry and radio science data have been used to create topography and gravity maps of the northern hemisphere (Smith et al. 2012). Figure 3.4 (Sun et al. 2012) is a topography map for the northern hemisphere. MESSENGER's elliptical orbit and high northern latitude periapsis precluded the capability to create a global topography map at this scale. A gravity anomaly map is shown in Chap. 2, for comparison. The implications for interior structure, requiring the use of both topography and gravity maps, were discussed in Chap. 2. Some global and regional implications for surface features are discussed here. Like the Moon, Mercury appears to have global-scale asymmetries in crustal thickness. The geoid indicates greater thickness at the two hot poles, as described above, as well as at lower latitudes, and is thinner at high latitudes, including the extensive northern plains region and at larger impact basins. Like the Moon, Mercury has large mascon-like anomalies associated with Caloris and several other large basins, as discussed above.

3.6 Tectonic Features and Historical Tectonism

By providing the first truly global maps, the MESSENGER mission has allowed more detailed understanding of Mercury's tectonic history on a regional basis, and refined our understanding of the influence of the late heavy bombardment, but our overall understanding of the major episodes has remained the same (Clark 2007).

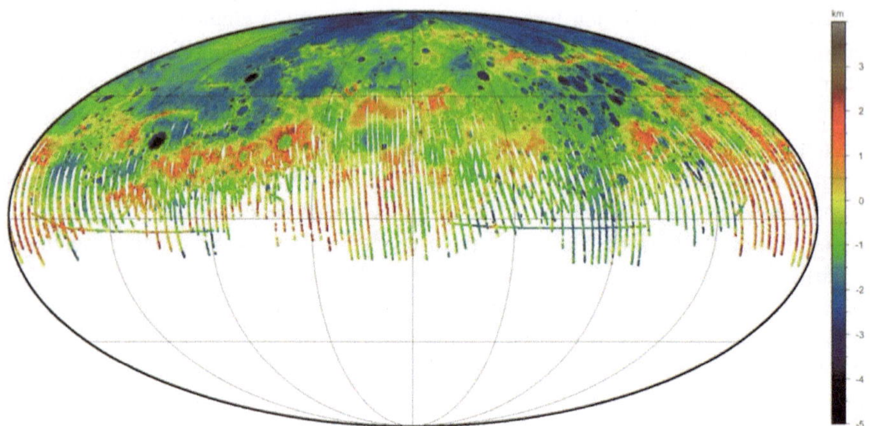

Fig. 3.4 MESSENGER MLA-derived topography map. (Sun et al. SPIE Newsroom, Fig. 3, http://dx.doi.org/DOI#10.1117/2.1201210.004489, 2012. © 2012 Society of Photo Optical Instrumentation Engineers)

Lobate scarps, surface expressions of large thrust faults of distinctive form, are confirmed to be the most globally distributed and common tectonic features. They were created during the era of solid core formation and consolidation that resulted in crustal shrinking on a global scale. Watters and coworkers (2009) ascertained, even early in the MESSENGER mission, a more spatially and temporally distributed network of scarps than previously assumed, indicating that earlier estimates of global contraction were too low and that solid core formation occurred over a longer period of time and may in fact still be occurring. Although they transect older features, scarps also show great complexity, changing in orientation, trend, dip, and degree of segmentation along their lengths, bifurcating or forming parallel ridges, and transforming periodically from high relief ridges into wrinkle ridges around basins and back into high relief escarpments (Egea-Gonzalez et al. 2012; Ruiz et al. 2012; Freed et al. 2012). Not only does this variability make estimating the extent of crustal contraction difficult. Such geomorphology indicates underlying structural effects from earlier episodes of global scale tectonism (at the very least, initial tidal despin resulting in a network of orthogonal extensional strike slip faults possibly followed by molten core formation and crustal expansion) (Clark, 2007) as well as earlier impact events that may have reactivated earlier tectonic features. Examples are shown in Fig. 3.5 (Ruiz et al. 2012).

Graben and wrinkle ridges are commonly associated with impact basins, as they are on the Moon. Freed and coworkers (2012) found that wrinkle ridges are apparently compressional features resulting from interior contraction that overlie crater rim/rings, indicating structural changes by partially or completely buried impact basins. These basins also have interior networks of grabens and ridges. They interpreted the extensional ridge and graben features to result from conservation of the area during cooling of thick flood lavas within craters. Wider floors are associated with thicker cooling units. Though Freed and coworkers (2012) can't explain concentric grabens

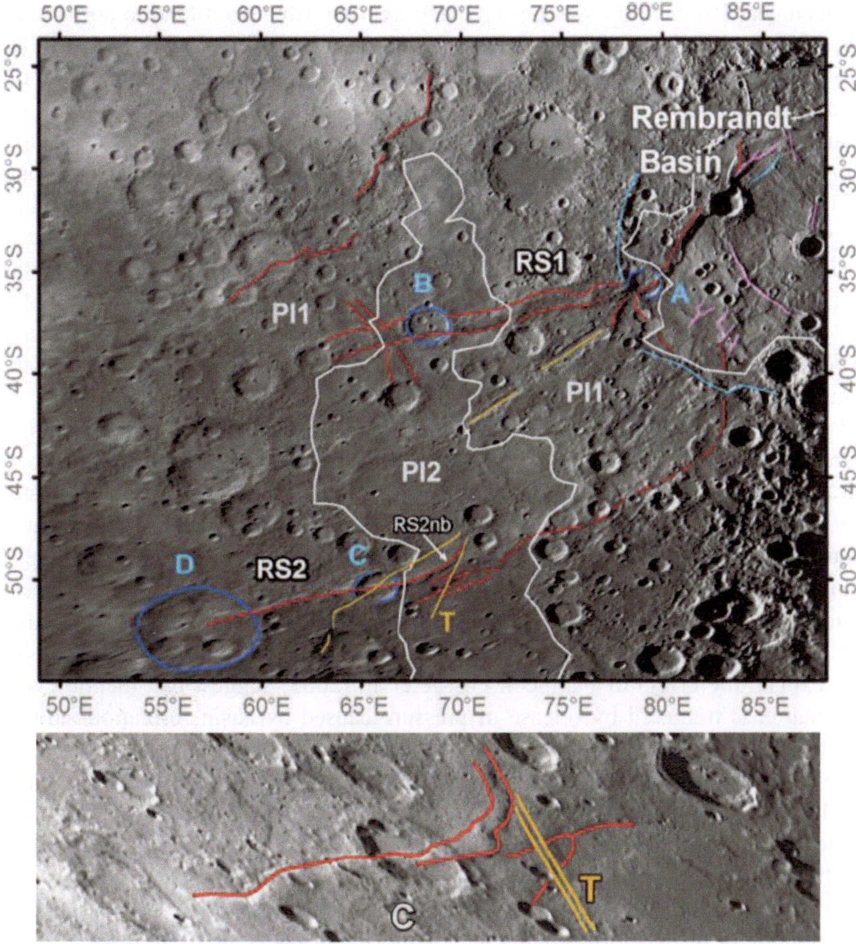

Fig. 3.5 Scarps (in *red*) including RS1 and RS2 exhibiting evidence of control by previous structures in intercrater plains (PI1) and heavily cratered terrain (PI2). Scarps, interpreted to be thrust faults, transform into wrinkle ridge-like features (*pink*), cross deformed impact craters (*A–D*), and intersect with troughs (*tan*) interpreted to be extensional faults or coalesced secondaries, including *T*, as described in the text. (Reprinted from Icarus, p. 219, Ruiz et al. 2012, Fig. 2 (*top*) and 3 (*bottom*), © 2012, with permission from Elsevier)

within impact basins, they may occur due to reactivation of a pre-crustal contraction orthogonal network of extensional features. The combination of high volume, low viscosity lava flows on top of already thick underlying lava deposits in interior craters of large impact basins creates a range of tectonic landforms not seen on the Moon or elsewhere in the Solar System (Freed et al. 2012).

Tectonic features associated with Caloris Basin have been extensively studied (e.g., (Basilevsky et al. 2011). Structural features of Caloris include: (1) the radial graben near Pantheon Fossae basin center, trending from azimuthal to north-south as a function of distance from center; (2) sinuous circumferential graben in a ring 350–600 km from center; (3) circumferential ridges extending from circumferential graben to the edge of the basin. Basilevsky and coworkers (2011) interpreted the Pantheon Fossae structure to be extensional radial grabens on an embayed rise that formed earlier to nearby tens of kilometers long and kilometers wide compressional ridges and circumferential extensional graben of the same size, but contemporaneously with the Caloris Basin floor materials unit. They interpret such structures associated with the Caloris Basin floor as the surface expression of a mantle diapir analogous to Venusian astra/novae where graben formation results from radial dikes propagating from magmatic centers. This model implies that compressional stresses associated with the ridges developed over time after the formation of the Caloris interior plains had loaded the lithosphere and that earlier extensional radial features exerted structural control over the compressional features that formed later. On the other hand, because these compressional ridges 'cut' the radial graben, Watters and coworkers (2009) interpret the radial graben as older, a point contested by Basilevsky and coworkers (2011), who argue that graben must be older because faults can't propagate across a free surface. Watters and coworkers (2009) interpret sinuous, circumferential graben to be discontinuous segmented polygons and circumferential ridges analogous to wrinkle ridges that intersect radial and circumferential graben, with displacement suggesting that grabens predate ridge formation. To explain the origin of Pantheon Fossae et al. (2005) suggest that mantle diapir formation is triggered by release of pressure caused by basin formation-induced melting. The resulting local relaxation of the impact cavity would result in heat release during crystallization and differentiation of large pools of impact melt (Cintala and Grieve 1998). However, the Pantheon Fossae structure is the only one of its kind yet observed on Mercury, which calls into question the diapir-induced Venus astra structure analogy. Other models imply that radial graben of Pantheon Fossae resulted from nearby Apollodorus Crater formation (Freed et al. 2009) or from uplift doming of the basin floor without a diapir (Klimczak et al. 2010).

According to Basilevsky and coworkers (2011), the postulated extensional global trough and ridge system has been observed on a local scale in only a few areas, including Caloris, from the time it was discovered.

Klimczak and coworkers (2013) have also studied tectonic activity near and in the Caloris Basin features, including Pantheon Fossae, and their implications for the thickness and strength of subsurface structures by using displacement profiles and displacement-to-length scaling. Near the center radial graben are no deeper than 4 km, the thickness of the underlying disrupted mechanical layer, but the graben deepen further from the center as they form directly on the flood basalt plains, where scaling indicates mechanically strong material. Altimetry data indicates long wavelength (< 1,000 distance and > 2.5 km depth) topography. Apparently, crosscutting and superposition relationships among various tectonic, impact, and volcanic structures at this location, and by implication in several other places where long-wavelength topographic features are seen, indicate that such topographic variations occurred relatively late, after volcanic plains emplacement, faulting, and post-Caloris cratering.

Around the Caloris Basin, Klimczak and coworkers (2013) see evidence that the smooth plains were affected by later developing extensional and tectonic stress which formed graben, most common and with preferred radial orientation toward the center, and wrinkle ridges, occurring further out with mostly concentric orientations. Exposures of material with different spectral properties in ejecta and central peaks is exposing underlying stratigraphy, indicating the underlying high-reflectance red plains composition of the disrupted mechanical layer exposed in the center of Caloris is up to 4 km thick. Faults do not crosscut craters, indicating that these tectonic features formed relatively quickly after plains emplacement. Tilted crater floors indicate that long wavelength deformation occurred during and after cratering. Klimczak and coworkers (2013) speculate that the deformations are too small in scale to be the result of global crustal compression, but that they might result from more local mantle processes, possibly resulting from Caloris formation. Crests and troughs of the long wavelength topography appear to correlate to gravity spherical harmonic to degree and order 20, but graben and wrinkle ridge are preferentially oriented only in regard to Caloris.

3.7 Volcanic Features and Historical Volcanism

Mercury's surface is primarily volcanic in origin, with two out of three of the primary terrains (older intercrater plains and younger smooth plains) having a primarily flood basalt origin. The third, highly cratered terrain, is compositionally similar to intercrater plains, despite being dominated geomorphologically by impact features.

Volcanic units range in signature from high reflectance red plains (HRP) with greater spectral slopes to low reflectance blue plains (LBP) with greater spectral slopes (Denevi et al. 2009). As described by Denevi and coworkers (2009), HRP, primarily associated with smooth plains (SP), are thought to consist of low Fe basalt, indicating more recent, lower temperature formation and a lesser degree of partial melting; LBP, primarily associated with intercrater plains, consist of higher Mg/Si, higher Ca/Si, and lower Al/Si basalt, indicating less recent, higher temperature formation and a higher degree of partial melting. Intermediate compositions, such as that of Rudaki, are compositionally related to HRP by way of partial melting. The LBP composition of circum-Caloris plains is similar to that of the surrounding heavily cratered terrain predating the smooth plains, suggesting that a common higher temperature source region with a higher degree of partial melting is not limited to older terrains. Two large low reflectance units include the smooth plains north of Beethoven and northwest of Rembrandt. Volcanic vents seen on the edges of the younger plains are typically depressions surrounded by redder, higher reflectance haloes. Lack of evidence of vents within the intercrater and smooth plains units are thought to be a result of the burial of sources by extensive flooding with low viscosity lava (Hurwitz et al. 2013). The range of compositions of volcanic units on Mercury suggests source magmas have lower or similar densities to surrounding crustal rocks, reducing the likelihood that variations in crustal thickness influenced magmatic activity and the composition of volcanic rocks (Hurwitz et al. 2013). The compositional evidence that the older intercrater plains have a volcanic origin suggests that the uneven distribution in smooth plains may be due to differences resulting from lower degrees of partial melting at the later volcanic emplacement age.

The unit defined as smooth plains is primarily volcanic and covers 27% of the surface (Denevi et al. 2013). SP distribution is hemispherically asymmetrical, more heavily concentrated in the northern hemisphere, and although no simple relationship has been established between plains distribution and crustal thickness (which decreases with latitude) or radiogenic element abundance, the crust beneath the northern volcanic plains, a smooth plains unit, is thinner than average. Rimless irregular depressions, interpreted as volcanic vents, are typically found on the margins of smooth plains exposures. The three major smooth plains formations, smaller in extent than globally distributed intercrater plains, include the Caloris interior, circum-Caloris plains, and the northern volcanic plains, and the terrain is apparently non-existent around 0° east, but that might be an observation effect, as this area was not covered by *Mariner 10* observations and was covered at low solar incidence angle only by MESSENGER (Denevi et al. 2013). Patches of smooth plains apparently formed as recently as 1 billion years ago. Spectrally, most smooth plains are like the northern volcanic plains—Mg-rich alkali basalt. A smaller fraction of the smooth plains unit has lower reflectance and a shallower spectral slope, implying an ultramafic composition and the higher temperature and higher degree of partial melting associated with the older, intercrater plains, and thus also implying that such conditions persisted in source regions during at least part of the later episode of SP formation (Denevi et al. 2013).

Smooth plains are typically very gently rolling, although knobbier terrain is not unknown (Denevi et al. 2013). SP are always sparsely cratered, indicating a narrower range of younger ages, like lunar maria. Caloris and the surrounding northern plains have a long wavelength topographic signature, indicating post-emplacement deformation. Crater densities are slightly higher in the interior of Caloris than to its east and west, and slightly lower in the interior than to the south of Caloris. Intercrater plains in contrast have an abundance of secondary craters, more gradational boundaries, and are somewhat rougher. According to Denevi and coworkers (2013), typical smooth plains exhibit effusive volcanism with characteristic lunar mare-type flooding, sharp boundaries, and embayment of older features, compressional wrinkle ridges, and level to gentle slope. However, knobby, hummocky (Odin-type) plains can be found around Caloris. Odin-type plains morphology and stratigraphy imply impact origin as Caloris ejecta, as do the variegated and diffuse boundaries of circum-Caloris plains. This interpretation conflicts with a younger post-Caloris impact age that would be implied by crater size frequency. If this is a volcanic unit, more effusive mafic deposits are implied. Volcanic landforms associated with Caloris also include small shield volcanic and volcanic vents. SP units with such landforms are also observed far from large basins.

After studying examples of lava channels on Mercury, Hurwitz and coworkers (2013) concluded that the morphology of plains on Mercury is similar to lunar mare plains morphology but lacking in constructional and erosional features, suggesting higher effusion rates of lower viscosity lava consistent with wide channels (Fig. 3.6). Four such channels (tens of kilometers wide by hundreds of kilometers

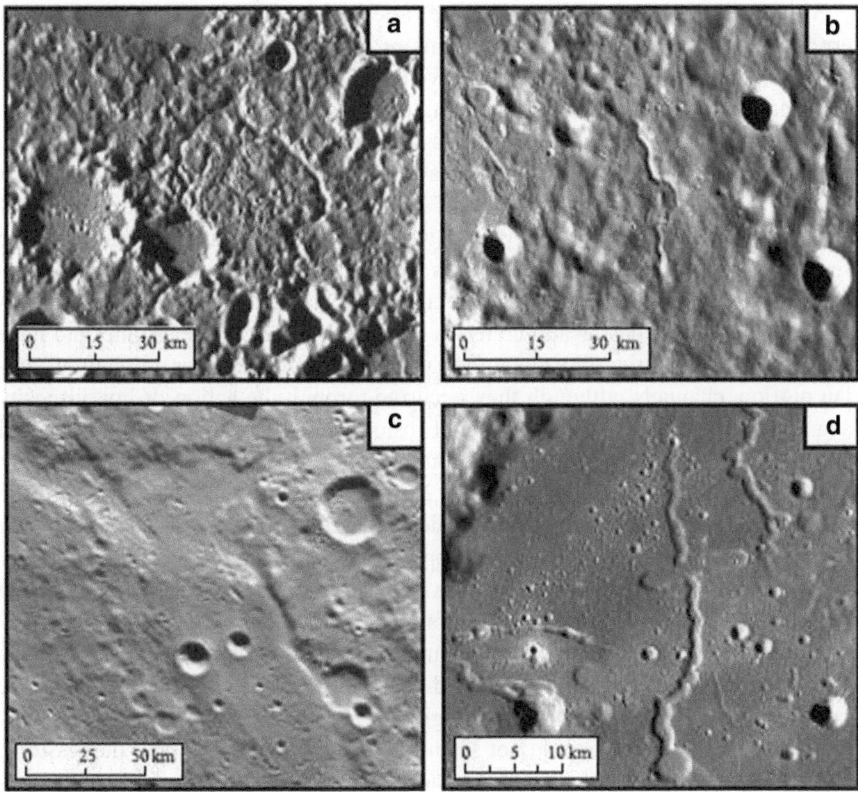

Fig. 3.6 Comparison of lava channels on Mercury and the Moon. **a** Mercury 25° S, 328° E; **b** Moon northern rim Imbrium basin; **c** Mercury 72° S, 167° E; **d** Moon Rimae Hanel and Telemann east of Harbinger Mountains. *a* and *b* show lack of source vents and termination deposits. *c* and *d* have nearly circular depressions upslope at potential sources. (Hurwitz et al. 2013, JGR Planets, Fig. 1, © 2013 American Geophysical Union)

long and hundreds of meters deep) radial to Caloris and with indicators, including grooves and streamlining parallel to valley walls, of lava outflow southward from the plains, formed on the edge of the northern volcanic plains, terminating in partially filled impact basins. These lavas, in an otherwise gradually sloping terrain with few higher relief structures, may provide the only places for lava to pool. Their two-step model involves mechanical erosion of the regolith followed by thermal erosion once a lower rigid layer is encountered (Hurwitz et al. 2013). Similar gravity would predict that thermal erosion rates for comparable scale lava channels would be the same on Mercury and Mars, but they are not; rates are higher on Mercury due to the higher temperature, lower viscosity, and the resulting higher effusivity or the komatiite-like ultramafic lava of Mercury. By contrast, narrower channels (tens of kilometers long by kilometers wide), lacking obvious eruptive sources (depressions associated with volcanic vents) and associated with impact structures, are thought to form from impact melt carving narrower channels into impact ejecta.

Although Mercury doesn't possess the variety of volcanic landforms of the other terrestrial planets, surface flow landforms of the high northern latitudes do have analogs on Earth, the Moon, Mars, and Venus (Byrne et al. 2013), including broad channels with streamlined kipukas cutting through older intercrater plains and narrower, more sinuous channels, and depressions, indicating large volumes of flowing, eroding low viscosity, high temperature lava. These streamlined features are thought to be the remnants of impact structures—central peaks, peak rings, rims, and remnants of sculpted terrain. In some places, where the broad channels exhibit furrowed floors, the surrounding intercrater plains show a relative decrease in elevation, as well as a smoother, more hummocky texture. The broad channels, which interconnect flooded impact basins and craters (e.g., Cahokia, Angkor, Paestum) and have no indications of stratigraphy along their walls, are particularly reminiscent of Martian outflow channels, although they occurred through the action of low viscosity lava and not water (Byrne et al. 2013). Erosional processes that formed the broad channels may have been amplified through the use of preexisting depressions of impact or tectonic origin (Fassett et al. 2009) as follows. The proximity of these assemblages to volcanic plains could indicate that they may have escaped complete inundation only because of a waning supply of lava. The smooth plains, effectively burying most preexisting structures, have lobate margins appearing as lava flow fronts, are spectrally homogeneous, but are distinctively inhomogeneous with surrounding older terrain. The narrow channels, on the other hand, may be analogous to lunar sinuous rilles, but evidence of distinct sources is lacking. They may have formed by lava escaping downhill through topographic lows in surrounding plains. Unlike broad channels, they do not seem to follow preexisting linear depressions, indicating entirely thermal erosion. Erosion via impact would not produce the evident spatial distribution or flux. Also, the apparent coalesced depressions along the routes of the broad channels appear, on the basis of their lack of raised rims, to be volcanic in nature. Alternatively, they could be overlapping impact craters acting as lava ponds drained by subsurface lava tubes, their rims erased by heavy lava flows.

Evidence of northern latitude volcanism revealed large expanses of smooth, contiguous flood plains with distinctive spectral features, indicative of high temperature, low viscosity lavas (Head et al. 2011). The extent of crater burial indicates lava thicknesses of greater than 1 km, and crater density ages indicate multi-phased emplacement during and following the late heavy bombardment and in fact subsequent to Caloris Basin formation. These plains are comparable to the smooth plains of the Caloris region, and apparently, like the Caloris plains, unrelated to underlying basin ages and thus unrelated to basin formation, unlike lunar mare. Ghost craters are frequently observed. Wrinkle ridges are also frequently seen, but with no clear association to underlying basins. Evidence for volcanic vents is lacking, as on the Moon. However, Head and coworkers (2011) suspect that evidence for vents may be seen by more thorough examination of the lobate margins of smooth plains, the youngest volcanic terrain, or at apparent points of origin of lava-carved channels. This supports the model of widespread partial melting of the mantle during and following the late heavy bombardment. The northern volcanic plains appear to be most analogous to the irregular Oceanus Procellarum.

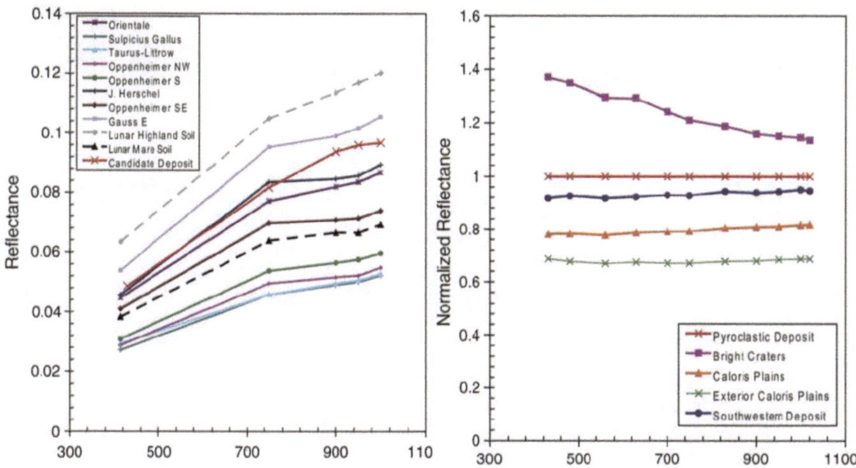

Fig. 3.7 Comparison of spectra for Mercury pyroclastics (from averaged individual spectra taken from vent area) to lunar pyroclastic, highland and maria spectra (Pieters and Tompkins 1999) (*left*) and to spectra from other Mercurian terrains (*right*). Mercury spectra lack a distinct 1 μm feature. (Reprinted from Earth and Planetary Science Letters, p. 285, Kerber et al. Fig. 3, *left*, and 4, *right*. © 2009, with permission from Elsevier)

MESSENGER images provide evidence for a variety of volcanic styles beyond flood volcanism (Kerber et al. 2009, 2011), including rimless and irregular pits, vent-like depressions, nearly circular positive relief features, and associated ballistically emplaced deposits known as pyroclastics. Spectra of likely pyroclastic deposits of the Moon and Mercury are compared in Fig. 3.7. Evidence for pyroclastics are typically found on central peaks and ring structures, as on the Moon, and not fractures in crater floors, but this could be the result of burial by later lava flows. Evidence of a shield volcano southwest of Caloris include central irregular depressions and diffuse-bordered bright deposits of what appear to be pyroclastics erupted at speeds of hundreds of meters per second based on the size of the deposit (tens of kilometers across) (Kerber et al. 2009, 2011). Volatile concentrations of several thousands of ppm (in CO, CO_2, SO_2, H_2O, H_2S), comparable to those of Hawaiian shield volcano eruptions, are required to achieve such a distribution. The presence of such volatiles and indications of a significant volatile budget in the mantle and crust of Mercury were unanticipated, but supported by bulk composition measurements derived from elemental abundance measurements.

3.8 Hollows

MESSENGER MASCS observations have led to the discovery of a new Mercury landform without analog. Hollows are bright, fresh appearing, irregular shaped, rounded edge, shallow deposits within impact craters. They appear to be bumpy,

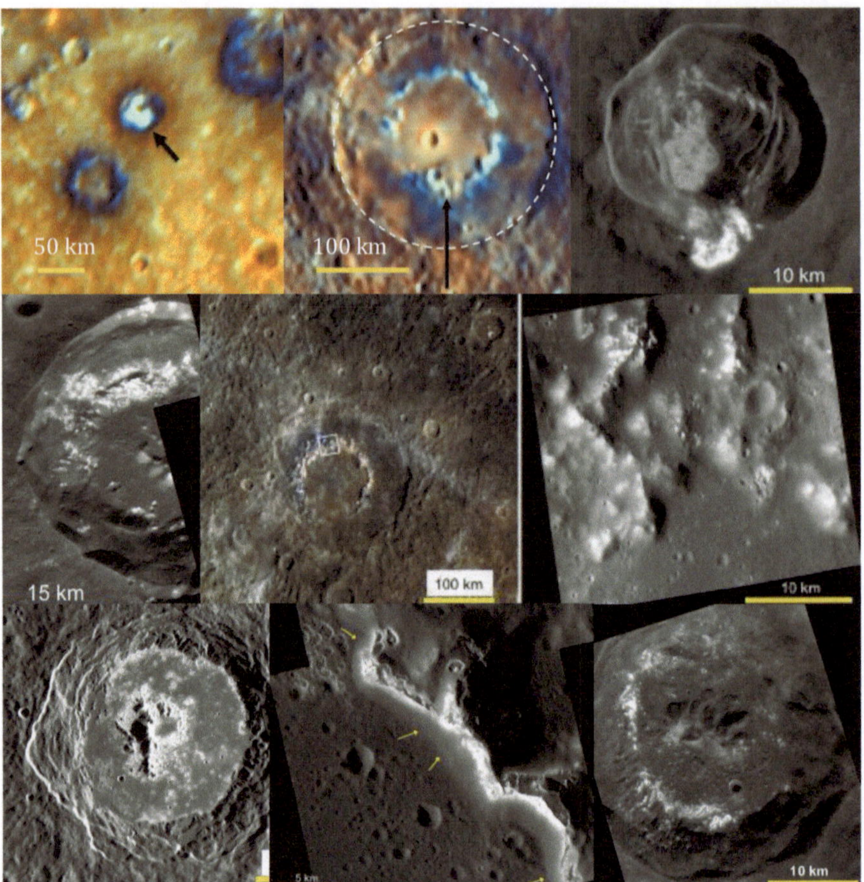

Fig. 3.8 Examples of hollows (Blewett et al. 2011): *Top*, *left* to *right*: crater at 42.5N, 154.7E filled with characteristic cyan color in principal-component enhanced-color image (Robinson et al. 2008; Denevi et al. 2009); raditladi with characteristic color associated with peak ring from same enhanced-color image; hollow signature on slump of wall in Kuiperian crater Dominici. *Middle*, *left* to *right*: crater at 50.8N, 319.9 E with signature on terrace floor; Vivaldi Basin hollow signature around peak ring; detail of Vivaldi peak ring. *Bottom*, *left* to *right*: Warhol Crater with bright hollows on smooth floor; detail of concentric hollows around peak ring of Raditladi Basin; crater at 37.8N, 321.2E with concentric hollow at floor/wall boundary. (Modified from examples in Blewett et al. 2011)

knobby, rimless depressions, excavated from depth in a process involving volatile loss, and they are tens of meters to a few kilometers in extent (Blewett et al. 2011, 2013) (Fig. 3.8). These occur on crater floors, walls, peaks, and rings. The volatile loss processes initially under consideration included sublimation, outgassing, pyroclastic activity, or some combination of these. The presence of features formed through volatile loss supports the model of a far more volatile-rich Mercury interior. The larger size of pyroclastic features, as well as their lack of occurrence in the interiors of impact craters, gives them a characteristically different signature from hollows. Fumarolic deposition and alteration of components unstable on the surface, after impact event exhumation, could be producing a 'Swiss cheese terrain' not unlike the Martian polar deposits (Blewett et al. 2011). These mechanisms would not preclude loss of unstable material such as sulfur through space weathering processes such as sputtering and impact vaporization.

Blewett and coworkers (2013) and Xiao and coworkers (2013) have provided in-depth characterization of hollows and associated features since their discovery. Their spectral signatures are distinctive, with their distribution primarily in hotter locations near the equator and on hot pole-facing slopes. Thus intense solar heating as well as interior processes are apparently contributing to their loss of volatiles. Hollows appear in a range of low reflectance units but are not formed directly in high reflectance plains (instead forming in excavated low reflectance material) and are rare in smooth plains. They are associated with craters ranging from simple bowl-shaped to multi-ring in size, Kuiperian to Calorian in age. Larger craters with larger hollows include Sander, Kertesz, de Graft, Tyagaraja, and Eminescu. These have etched features around topographic highs. Older, more subdued hollows are found in older craters, such as Abu Nuwas. Hollows on wall terraces and slumps are preferentially found on south-facing walls. Hollows on central peaks and rings are found on flat mountaintops and talus slopes, as in Raditladi and Vivaldi. In Raditladi, the talus slope material, potentially derived from the hollow, has buried downslope craters and this may in fact be an ongoing event. Hollows form in small, simple craters as well. The presence of a hollow-forming layer at a constant depth below the rim around much of a crater is an indication of exposure of a shallow stratum with a lithology susceptible to hollow formation. On the other hand, hollows have been found in close association with pyroclastic deposits in several locations, including the floor of Praxiteles and Tyagaraja, Lermontov, and Scarlatti craters.

Why are hollows bright? Blewett and coworkers (2013) have proposed alteration products (during sequestration), vapor-deposited coatings, and destruction of darkening agents. Microphase or nanophase sulfide materials are thought to contribute to low reflectance (potentially in dark 'halos' as described below). The high reflectivity of hollows could be due to efficient destruction of such a darkening agent, or to physical differences such as smaller particle size, more complex grain surface morphology, or greater density of multiple scattering centers resulting from the formation mechanism.

Blewett and coworkers (2013) proposed that hollows originate as magmatic gases and fumarolic minerals initially exposed at the surface condense during Mercury's long, cold nights. They are then buried by subsequent volcanic activity and sequestered, causing modification of surrounding material until exposed through an impact event, initiating rapid sublimation and formation of a collapse pit. Impact melt differentiation could be producing lithologies rich in volatiles.

S in sulfides, Cl (chlorine), and even Na in plagioclase have been proposed as the volatiles implicated in hollow formation. Certainly subsequent space weathering processes contribute to the loss of volatile-bearing phases. A two-stage process of iron sulfide vaporization followed by ion bombardment of asteroids has been demonstrated in the laboratory to result in sulfur loss (Loeffler et al. 2009), and these processes could easily be happening on Mercury's surface, although S escape would be more difficult on higher gravity Mercury.

Xiao and coworkers (2013) found 'dark spots,' small, isolated units with the lowest average reflectance observed on Mercury. Dark spots exhibit a red slope, like the surface of Mercury, and may vary in reflectance by a factor of 2. Two types of atypically low reflectance unit (Xiao et al. 2013) include the circum-Caloris low reflectance blue plains (LBP) and the lowest average but variable reflectance LRM associated with the dark spots. Spectral signatures of LBP, associated with older volcanic units, cluster tightly; the spectral signatures of somewhat lower reflectance LRM associated with impact craters show more variability; and the LRM spots not associated with impact craters exhibit the lowest reflectivity and more variability and overlap with LRM associated with impact craters. The darkest spots, when observed at sufficient resolution, are thin with diffuse margins and feature hollows in the center, but, according to Xiao and coworkers (2013), only 30% of the hollows are surrounded by dark spots, or 'dark halos.' All hollows are irregularly shaped and rimless and have higher reflectance than surroundings. The 34 dark halos are globally distributed but not found on HRP and are considered to be the youngest endogenic features, best preserved in the freshest craters and even post-dating young craters, and possibly forming today. The other 48 small LRM areas, typically higher in reflectivity than dark halos, are found in impact craters on all types of terrain, including HRP, and are thought to be excavated LRM. Based on spectral characteristics LRM dark spots have a distinctively different composition from the higher reflectance and tightly clustered LBP material. Xiao and coworkers (2013) have proposed that dark halos contain a combination of sulfur and sulfide deposits, outgassed during hollow formation and then eroded or modified by 'gardening' on timescales much shorter than the hollows themselves. Dark haloes have a distinctively different composition from reddish deposits identified as pyroclastics and may have a different mode of emplacement, potentially being dispersed ballistically as volatiles up to several kilometers away, leaving a void and resulting in formation of a collapse pit with a bright floor at the point of release. As the dark halo disappears, a bright halo forms and grows over time. The composition and variability of the remaining LRM spots could result from impact excavation of crust with vertical heterogeneity in sulfide abundance due to different degrees of mixing during crust formation.

3.9 Polar Deposits

Measurements from the MESSENGER neutron spectrometer provided further evidence for the startling discovery of 'icy satellite' signatures in permanently shadowed craters at Mercury's poles based on ground-based radar observations made two decades ago (Slade et al. 1992). Focusing on the decrease in epithermal and fast neutron flux in these regions, Lawrence and coworkers (2013) confirmed the presence of a hydrogen-rich layer tens of cm thick beneath a surficial layer 10–30 cm thick with far lower hydrogen abundance. Epithermal flux is most affected by the presence of protons, but the fast neutron flux is more sensitive to variations in average atomic mass as a function of depth, and so can be used to determine vertical distribution of proton-bearing material (Lawrence et al. 2013). Observations from both datasets are best matched if the buried layer consists nearly pure water-ice, and the upper layer is $<25\,\%$ water-ice. Estimates made on this basis combined with estimates of the total area of permanent shadow from radar observations and assumptions of tens of meters thickness of the ice layer indicate a total water mass of 10^{13}–10^{15} kg, which is consistent with delivery by comets or volatile-rich asteroids. Burial of ice layers is also consistent with MLA and MDIS observations, suggesting that surficial water-ice would not be stable for most radar bright features and could contain complex hydrocarbons because the surficial layer is darker (at visible wavelengths) than surroundings for these features. When combined with these data, vertical and lateral mixing models, which suggest a burial rate of 0.43 cm/million years (Crider et al. 2005; Butler et al. 1993), would imply emplacement of water-ice in polar cold traps within the last 18–70 million years.

Chabot and coworkers (2012, 2013) have done a systematic study of radar bright features at Mercury's poles. These features have the distinctive polarization ratios of ice. In addition, sulfur is stable at higher temperatures than ice. If these features were sulfur, they would be larger and present in somewhat shallower and lower latitude craters, and in fact generate a cap within one degree of the pole (Butler 1997; Vasavada et al. 1999). At the south pole, 85 % of the radar bright features correspond to craters with permanent shadow, and most of those that don't are in Chao Meng Fu, where small-scale structures are smaller than the limit of resolution. At the north pole, all polar craters observed to have permanent shadows have radar bright features. Craters lacking radar signatures at the poles tend to be found at lower latitudes and nearer 'hot poles.' Further south, radar bright craters are found nearer the cold poles, and models suggest the need for an insulating layer of regolith. Within 10° of the poles, almost all the craters <10 km in diameter have radar bright deposits. Several permanently shadowed craters at the north pole are large enough to contain a long-lived thermally stable environment for water-ice. In the north, radar bright craters are found down to 66° south but may not be stable even with an insulating layer. Thermal modeling of radar bright lower latitude craters imply emplacement of <50 million years (Chabot et al. 2013). Differences in distribution between north and south poles are thought to be due to less complete radar viewing of the south pole at a different geometry. The location of the coldest area in the permanent shadow varies, depending on the crater size, depth, shape, orientation, and resulting thermal profiles. Long wavelength topography, such as broad rises, could affect the location of the coldest areas. Altered dielectric properties of silicates at low temperatures have also been postulated to produce high radar backscatter (Starukhina 2001); however, if this were the case, a correlation between radar brightness and permanently shadowed craters would be expected on the Moon, which is colder, but such a correlation is not observed.

MESSENGER MLA observed bright and dark deposits in permanently shadowed craters at 1,064 nm (Neumann et al. 2013). Deposits are concentrated on poleward-facing slopes and correlated with radar bright features. Correlation with modeled surface temperatures lends credence to the hypothesis that optically bright spots are surface exposed water-ice, and dark spots a complex organic layer acting as a thermal insulator over surface ice. Larger dark areas are found in areas of low illumination but not necessarily permanent shadow, are more widespread than radar bright spots, and, when found with radar bright spots, enclose the radar bright material. This material may in fact form a radar-transparent layer on radar bright spots. Optically bright spots are associated with radar bright spots, surrounded by dark spots, in the highest latitude permanently shadowed craters. All of these associations indicate water-ice deposition in geologically recent times, possibly even at present, and rates of formation greater than impact 'gardening.'

Paige and coworkers (2013) have developed thermal stability models for volatiles on Mercury described here. The biannual average temperature, a function of latitude and longitude on Mercury, is a close approximation of the subsurface temperature at or below the depth of the diurnal temperature wave, from 0.5 m for ice-free regolith to several meters for ice-rich regolith. This information can be used to determine subsurface stability for water-ice and other volatiles. Volatiles have very different stability profiles. 1 mm water-ice will sublimate in 1 billion years between 100 K and 115 K, a reasonable temperature range and sublimation rate for observed water-ice deposits at the poles and indicative of their longevity. Darker organic material would be even more thermally stable and probably form as sublimation lags while water-ice sublimates and retreats to a more thermally stable regime. Lag formation requires episodic and coincident deposition of water and non-water contaminants at rates greater than sublimation rates. The formation of macromolecular carbonaceous material from simple organics in a high-energy charged particle environment is not inconsistent with lunar surface observations made by the LCROSS mission. A greater proportion of the less volatile organic components would result in longer duration water-ice deposits, which would require less frequent replenishment by external source, such as comets or volatile-rich asteroids, in order to persist.

3.10 Impact Features and Bombardment History

Mercury's gravity is 2.3 times that of the Moon, resulting in significantly different scaling in the size and complexity of its craters. A number of coworkers have attempted to establish a morphometric framework for Mercury's impact features as a means to understanding crater formation and the transition from simple to complex craters (Baker et al. 2011; Baker and Head 2013; Schon et al. 2011). Mercury has the largest population and greatest density of peak-ring basins and protobasins, and a type of basin not previously identified, the ringed peak-cluster basin, as part of a continuum of basin morphologies (Baker et al. 2011). Although many of the larger basins on Mercury may be obscured by the planet's early extensive flood volcanism, crater size distributions between 100 and 300 km in diameter are similar for the Moon and Mercury. The onset rim-crest diameter for peak-ring basins is

126 km, intermediate between the onset diameters of the Moon and other terrestrial planets, and is much lower on Mercury than the Moon by a factor of 5 (Baker and Head 2013). Mars has fewer peak-ring basins due to extensive resurfacing, but the mean and onset diameters for these basins are similar on the two planets, which have similar gravity. On the other hand, some workers have seen indications that the projectiles bombarding Mercury may have lower mean densities, due to a high proportion of comets (Barnouin et al. 2012). Baker and Head (2013) have found evidence for much more overlap in diameter between different crater types on Mercury, resulting from a broader distribution of impact velocities. As central structures are impact-velocity dependent, this observation implies broader distribution velocities to account for more variation in morphologies for given diameters (Baker and Head 2013). Peak-ring basins have greater prevalence, smaller onset diameter, and a narrower range of diameters compared to the Moon. Single peak craters are shallower on Mercury than the Moon, which may be a gravitational effect or a result of different pre-impact topography and underlying stratigraphy.

Figure 3.9 (Baker et al. 2011) illustrates crater classification on the basis of features described here, systematically increasing in diameter: (1) simple craters, (2) modified simple craters, (3) immature complex craters, (4) mature complex craters, (5) protobasins, (6) peak-ring (two-ring) basins, and (7) multi-ring basins. Peak-ring styles vary on Mercury, ranging from partial arcs to escarpments to circular wrinkle ridges enclosing smooth plains to hybrids. Protobasins have more subdued and less complete rings, trends in rim-crest diameter showing a size-scaling process. Similar trends are observed on other bodies. Mercury has the largest density of protobasins and peak-ring basins in the Solar System, consistent with the high mean velocities of its impacts.

Baker and others (Baker et al. 2011; Baker and Head 2013) have studied the systematics of impact feature morphology within each crater class and transitions between single peak, peak-ring, and multi-ring basins. Structural features scale with outer diameter within each class. All interior peaks are > 1 km below the outer rim. In single peak craters, central peaks scale with diameter. An abrupt decrease in crater depth and in floor area to interior area ratios occurs in the transition from complex craters with central peaks to basins with peak rings. Mercury, like the Moon, has an abrupt decrease in crater depth and an abrupt end to the continuous increase in central peak dimensions, with crater diameter in the transition between crater and basin. The change in morphology implies a very different process for peak-ring formation. At the transition between complex and peak-ring crater, a slight reduction in total peak volume occurs, and then peak volume increases as the diameter of the peak-ring basin increases.

Trends in floor and interior area features are correlated with trends in impact melt production and retention (Baker and Head 2013). Although impact melt volume scales nonlinearly with crater diameter, floor radius increases relative to outer rim diameter in larger basins, apparently indicating increased impact melt production and retention. The relationships between peak-ring basins and protobasins are illustrated by the agreement between power law fits of ring diameter to rim-crest diameter ratios, and between melt cavity dimension and crater diameter for peak-ring

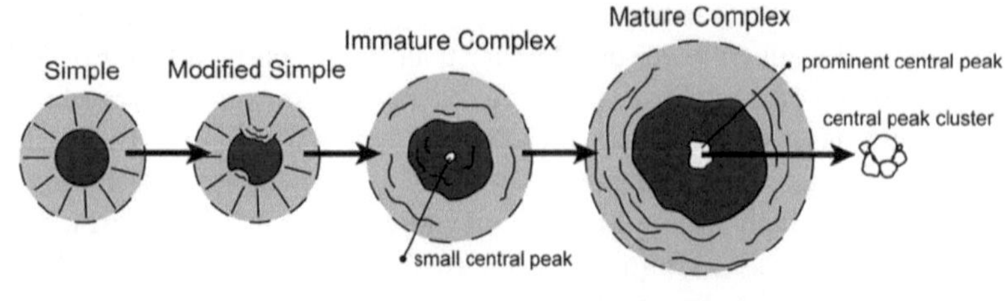

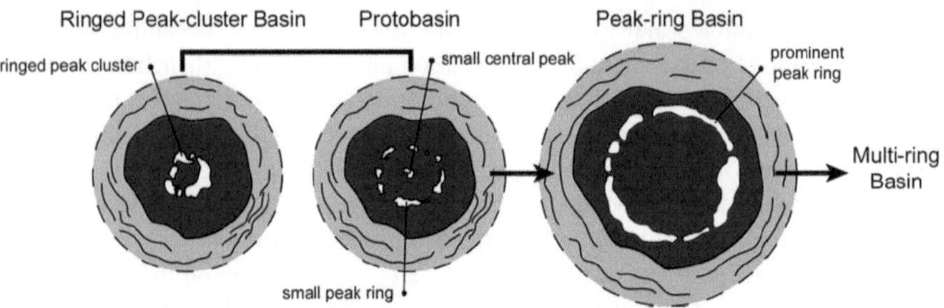

Fig. 3.9 Schematic progression of crater morphologies as a function of diameter as described in the text. (Reprinted from Planetary and space science, p. 59, Baker et al. 2011, Fig. 1, © 2011, with permission from Elsevier)

basins and protobasins (Fig. 3.9). The interior is at a lower elevation than the peak ring, which surrounds a cavity-like depression. Protobasins, which retain central peaks, result from more complex processes than previously predicted. Central peak or peak-ring area increases as the rim diameter increases, although the peak ring may decrease in height as the protobasin diameter increases.

Baker and coworkers (2011) and Baker and Head (2013) evaluate two formation models of larger impact features ranging from complex craters to peak-ring basins: the hydrodynamic collapse model (Melosh 1989) and the nested melt-cavity model (Cintala and Grieve 1998) (Fig. 3.10). The dynamic collapse model predicts fluid-like gravitational collapse of an over-heightened central peak. The target weakens and collapses, rebounds, and the outward wall collapses to form an unstable central uplift. In peak-ring basins, the downward and outward collapse overrides the collapse of the transient inner cavity wall, and the central peak material flows outward to form a peak ring. However, this model does not predict the observed reduction in central peak height or dimensions. The nested cavity model predicts non-proportional growth in impact melt volume with increasing basin size plus an increase in depth of melting as the depth of the transient cavity increases, weakening the central uplift. The most displaced and shocked material from the central peaks form. The depth of melting is not sufficient in complex craters to modify the initial uplifted morphology. In peak-ring basins, melt material streams downward and outward during transient cavity growth. During the following stage of rebound and collapse, the inner melt region translates inward and upward, but is deep enough to retard central peak formation, leaving the periphery of the central melt cavity the most prominent feature.

For the smaller peak-ring basins, a tiny central peak can still be observed in the center. Observations of interior depressions, and the relationship between melt production

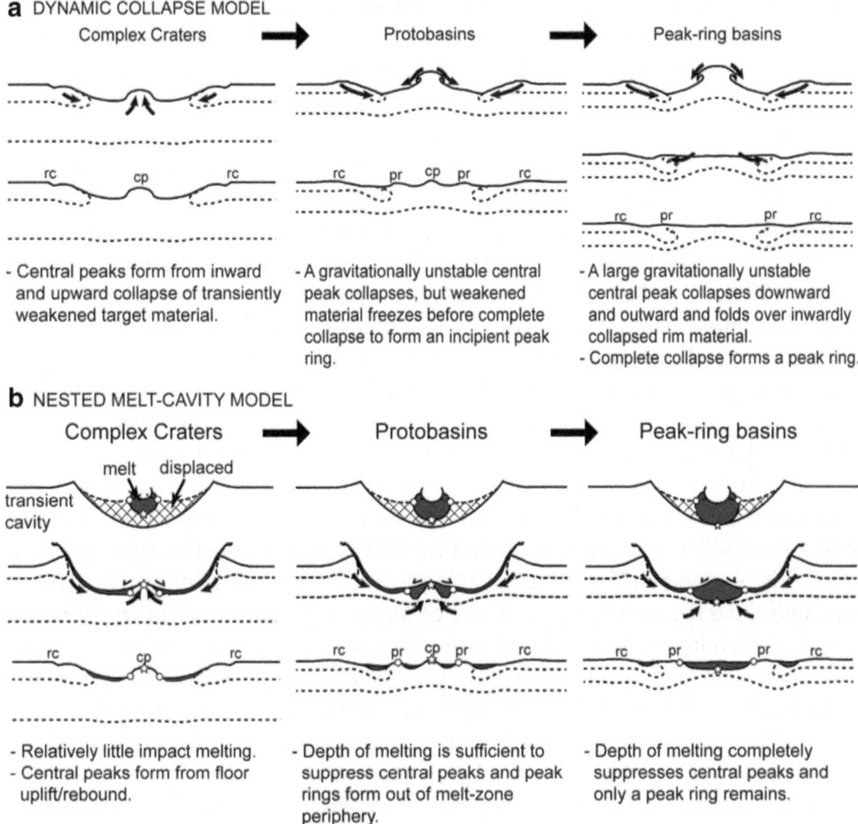

a DYNAMIC COLLAPSE MODEL
Complex Craters
Protobasins
Peak-ring basins
rc
cp
rc
rc
pr
cp
pr
rc
rc
pr
pr
rc
- Central peaks form from inward and upward collapse of transiently weakened target material.
- A gravitationally unstable central peak collapses, but weakened material freezes before complete collapse to form an incipient peak ring.
- A large gravitationally unstable central peak collapses downward and outward and folds over inwardly collapsed rim material.
- Complete collapse forms a peak ring.
b NESTED MELT-CAVITY MODEL
Complex Craters
Protobasins
Peak-ring basins
melt
displaced
transient cavity
rc
cp
rc
rc
pr
cp
pr
rc
rc
pr
pr
rc
- Relatively little impact melting.
- Central peaks form from floor uplift/rebound.
- Depth of melting is sufficient to suppress central peaks and peak rings form out of melt-zone periphery.
- Depth of melting completely suppresses central peaks and only a peak ring remains.

Fig. 3.10 Comparison of dynamic collapse and nested melt-cavity models as discussed in text. (Reprinted from Planetary and space science, p. 86, Baker and Head 2013, Fig. 12, © 2013, with permission from Elsevier)

and morphology, are more consistent with the nested cavity model, and the evolution of interior morphologies between protobasins to peak-ring basins is consistent with the progression of diameter-dependent morphology predicted by this model (Baker et al. 2011; Baker and Head 2013). The uplifted periphery of the melt cavity emerges as the dominant interior morphology, except for the ringed peak cluster basins, which are apparently not on the continuum but are an 'alternative form.' Ringed peak cluster basins follow the trend for central peak basal diameters in complex craters, which suggests that these craters form in an uplift-dominated regime, where impact melting is deep enough to penetrate the central impact structure. The array of separate peak elements appears to be made up of 'fragments' of a central peak. Both impact melt volume and rim diameter scale nonlinearly with ring/rim diameter for protobasins and peak-ring basins. Trends in floor and interior area are mostly due to differences in impact melt production and retention (Baker and Head 2013).

Barnouin and coworkers (2012) have also studied the systematics of impact feature morphology on Mercury in comparison with other bodies. Laser altimetry observations indicate that craters larger than 10 km (resolution limit due to 800 m spacing of 20–30 m footprints) are shallower on Mercury than *Mariner 10* implied, although the diameter of transition from simple to complex craters is consistent with *Mariner 10* observations. Photogeological data from MDIS indicate significant modification, allowing estimates of the volume of infilling material. Even small, simple craters are observed to be shallower than predicted on the basis of crater formation models (Pike 1988).

Schon and coworkers (2011) provide a model that illustrates the non-linear growth of the volume of impact melt with increasing crater volume (Cintala and Grieve 1998). A central peak is derived from uplift of material in the central melt zone. The impact creates a melt cavity that modifies the central uplift zone. Peak rings are formed from uplift of areas offset from the central melt zone. Formation of a central peak is inhibited for peak-ring basins by the growth of the central melt cavity due to greater energy input from a larger impact event. The impact creates a more extensive and hotter melt cavity that dominates the central zone.

The distribution of impact basins on Mercury gives us a record of the impactor population in the inner Solar System. The non-uniform distribution is consistent with a primordial synchronous rotation for Mercury rather than its subsequent 3:2 spin:orbit resonance (Correia and Laskar 2012; Wieczorek et al. 2012). The researchers used the distribution and size of impact features to develop a collisional model of Mercury in its postulated earliest rotational state and were able to reproduce its later large impact and even induced spin:orbit resonance through secular evolution with 50% probability as well. This result appears to be robust to variations in the initial spin state as well as anticipated variations in dissipation and collision. They used larger craters (> 300 km) because they are less apt to be modified by later events and also represent the size of impactor that would have sufficient influence to induce changes in Mercury's rotational state. Correia and Laskar (2012) used two different impact populations to simulate the effect of an impact on the planet's spin evolution. Both models predicted the 3:2 spin:orbit state of the most probable outcome.

Neish and coworkers (2013) have compared radar (S-band) and optical observations for > 20 km young rayed craters on the Moon and Mercury. Their radar studies of lunar craters have indicated a systematic decrease in impact ejecta size as a maturing process, as debris is 'gardened' from constant bombardment once exposed at the surface, and a characteristic brightness signature when debris size is equal to or somewhat larger than the radar wavelength. This brightness is associated with the primary ejecta interior walls and downrange debris surge. For a given crater size, Mercury has more secondaries than the Moon. Rays are longer on Mercury, up to 4,500 km from the parent crater. The models predict smaller secondaries and shorter rays due to higher impact velocities (Neish et al. 2013). This inconsistency may be due to increased crustal strength of extensive volcanic rock, higher ejection velocity of secondaries, and ejection of more concentrated clusters. Secondary cratering plays a significant role in the formation of rays, which are both optically and radar 'bright,' indicating a dense secondary population. Radar brightness at this wavelength (12.5 cm) is consistent with deposition of blocky, immature ejecta, in the centimeter to decimeter size range, from secondaries or rocky, immature interior walls. All but

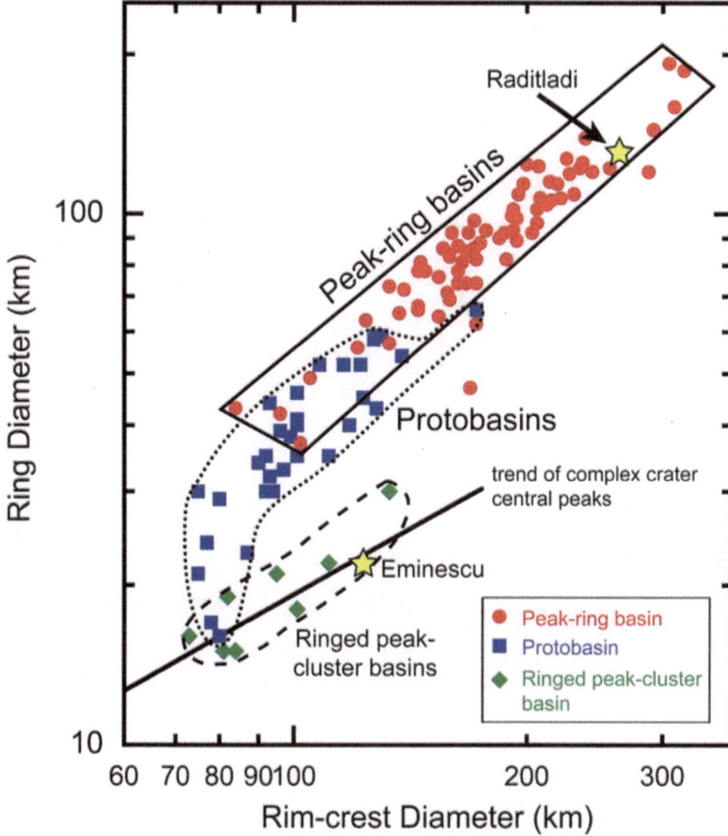

Fig. 3.11 Ring diameter versus crater rim-crest diameter for three basin forms. (Reprinted from Planetary and space science, p. 59, Schon et al. 2011, Fig. 3b, © 2011, with permission from Elsevier)

the youngest lunar rays are less radar bright at this wavelength, suggesting smaller clasts and a process influenced by differences in impact velocity, gravity, and target properties. Larger secondaries on Mercury would produce larger debris blocks that would require more time to degrade. On the other hand, Mercury's increased exposure to solar wind and micrometeorite bombardment combine to produce more rapid optical maturation, darkening, and apparent disappearance of rays on the planet.

3.11 Crater Ages and Geological History

Despite the fact that crater populations have been modified by volcanism and extensive secondary crater formation, the overall trends in crater populations (size vs. degradation versus density), as well as observable superposition, expose several

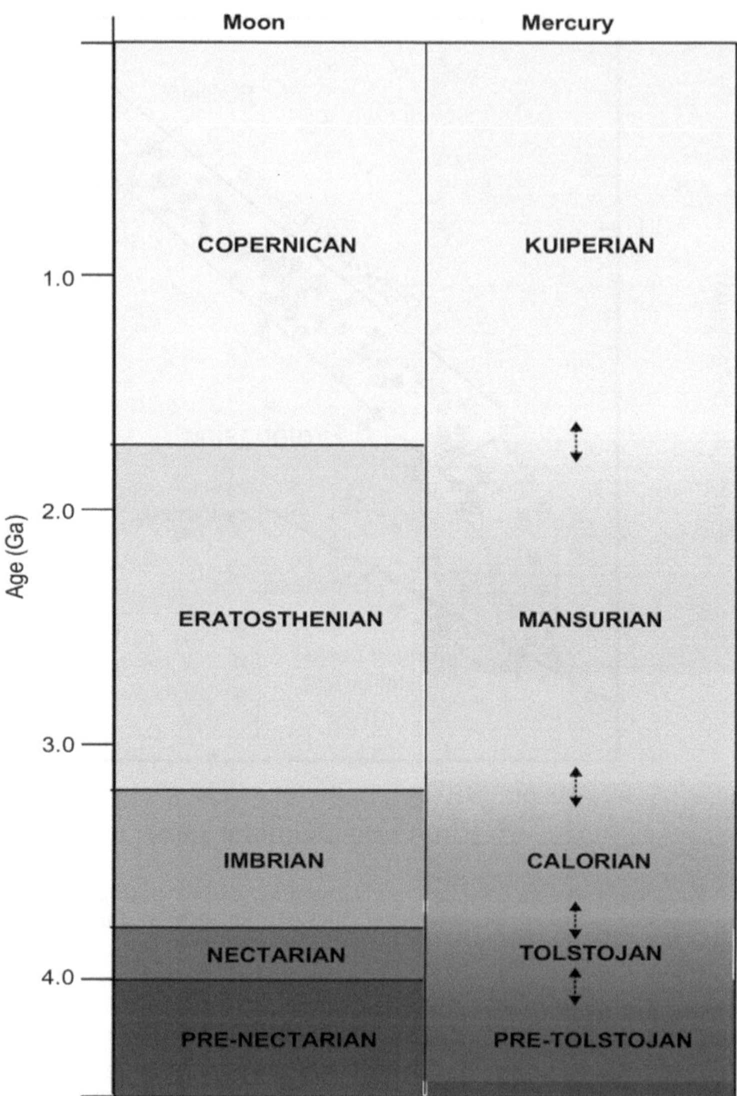

Fig. 3.12 Stratigraphic timescales on Mercury (Spudis and Guest 1988) considered roughly analogous (based on crater age dating) to the Moon (Wilhelms 1987). (Reprinted from Planetary and space science, p. 59, Schon et al. 2011, Fig. 7, © 2011, with permission from Elsevier)

major sequential episodes in that planet's history (Figs. 3.11 and 3.12) (Strom et al. 2011; Schon et al. 2011). The two major identifiable phases, associated with two major volcanic terrains, are intercrater plains formation and the later smooth plains formation (Pre-Tolstojan, Tolstojan, Calorian). Intercrater plains formed during the late heavy bombardment and were both considerably modified by bombardment and modifying of impact features by embayment and burial. Smooth plains formed

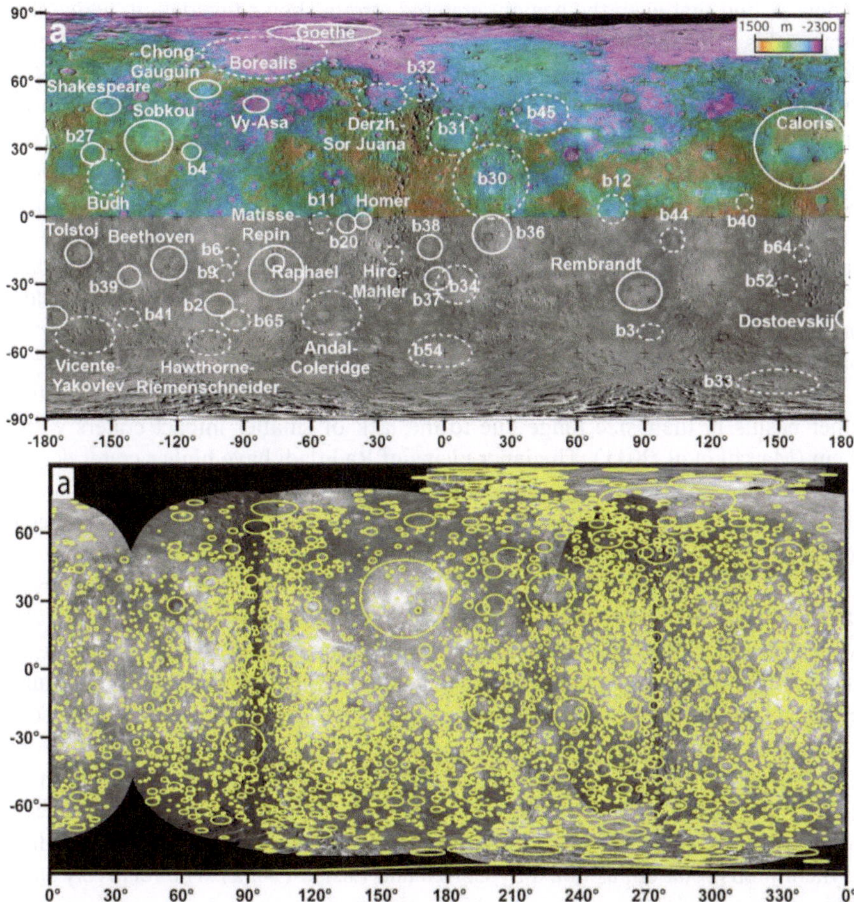

Fig. 3.13 Comparison of global crater density (>10 km) (*bottom*) and large (>300 km) basin distribution (*top*). *Top*: Fassett et al. 2012, JGR Planets, Fig. 3.1a, © 2012 American Geophysical Union. *Bottom*: Fassett et al. 2011, GRL, Fig. 1a (© 2011 American Geophysical Union)

interior and exterior to Caloris Basin (Calorian). Crater-size frequency studies indicate a younger population of impactors for the smooth plains overall; however, intercrater plains formation occurred over a long time period, and overlapped with smooth plains formation (Strom et al. 2011). The youngest smooth plains are considerably younger than the Caloris plains (Mansurian, Kuiperian). Crater population statistics indicate two major impactor populations: a larger, earlier one followed by a smaller, later one, just as on the Moon and Mars (Strom et al. 2011). Mercury does exhibit a pronounced regional variation in evidence for cratering, but craters are more likely to be obliterated in volcanic terrain.

Fassett and coworkers (2011, 2012), who also performed crater density studies (Fig. 3.13), observed a deficit of craters between 20 and 100 km as compared to the

Moon, noting that even the heavily cratered terrain shows this deficit, which they proposed is the result of volcanic resurfacing. Craters larger than 100 km and below 20 km show a similar distribution to lunar craters in that size range (Fassett et al. 2012), indicating a comparable experience of late heavy bombardment saturation cratering and a comparable recent cratering history. In fact, they found evidence that the Moon and Mercury had a similar rate of resurfacing after the late heavy bombardment. They also note that the smooth plains around Caloris are the youngest extensive and contiguous region on Mercury. The smooth plains appear to be considerably younger than the basins in which they are found. Herrick and coworkers (2011) interpreted the factor of 10 variation in areal crater density in volcanic plains on Mercury to indicate that plains resurfacing was occurring in large areas up to a billion years after late heavy bombardment ended.

Some basins, such as Raditladi and Rachmaninoff, appear to be younger than other basins in their size range due to the lack of smaller impact craters within them (Marchi et al. 2011). The inner plains of Raditladi have higher crater density than the surrounding annular units, giving a paradoxically older age. Marchi and coworkers workers (2011) propose that these differences result from differences in material properties of the substrate, indicating a layered and disrupted target.

Fassett and coworkers (2012) also studied the distribution of the largest (>300 km) basins on Mercury, and determined they were more degraded and with lower crater densities than comparable basins on the Moon (Fig. 3.13). This result is consistent with volcanic resurfacing or relaxation. In addition, the larger volume of impact melt thought to be generated on higher gravity Mercury could be obscuring the structures of multi-ring basins. The researchers also noted a non-uniform distribution of larger basins. Evidence for larger degraded basins was found in the heavily cratered terrain (possibly due to lack of complete burial). Fewer basins were observed in the eastern hemisphere, which they postulate was the result of asynchronous rotation, as noted by Correa and Laskar (2012). The relatively small dynamic range of topography in the high northern latitudes indicates a thin crust.

References

Baker, D., Head, J.: New morphometric measurements of craters and basins on Mercury and the Moon from MESSENGER and LRO altimetry and image data: An observational framework for evaluating models of peak-ring basin formation. Planet. Sp. Sci. **86**, 91–116 (2013)

Baker, D., Head, J., Schon, S., Ernst, C., Prockter, L., Murchie, S., Denevi, B., Solomon, S., Strom, R.: The transition form complex crater to peak-ring basin on Mercury": new observations form Messenger flyby data and constraints on basin formation models. Planet. Sp. Sci. **59**, 1932–1948 (2011)

Barnouin, O., Zuber, M., Smith, D., Neumann, G., Herrick, R., Chappelow, J., Murchie, S., Prockter, L.: The morphology of craters on Mercury: Results from MESSENGER flybys. Icarus. **219**, 414–427 (2012)

Basilevsky, A., Head, J., Fassett, C., Michael, G.: History of tectonic deformation in the interior plains of the Caloris basin, Mercury. Sol. Syst. Res. **45**(6), 471-497 (2011)

BepiColombo System and Technology Study Report, (2000). ESA-SCI (2000)

Blewett, D., Robinson, M., Denevi, B., Gillis-David, J., Head, J., Solomon, S., Holsclaw, G., McClintock, W.: Multispectral images of mercury from the first MESSENGER flyby: Analysis of global and regional color trends. EPSL. **285**, 272–282 (2009)

Blewett, D., Chabot, N., Denevi, B., Ernst, C., Head, J., Izenberg, N., Murchie, S., Solomon, S., Nittler, L., McCoy, T., Xiao, Z., Baker, D., Fassett, C., Braden, S., Oberst, J., Schoiten, F., Preusker, F., Hurwitz, D.: Hollows on Mercury: MESSENGER evidence for geologically recent volatile-related activity. Science. **333**, 1856–1857 (2011)

Blewett, D., Vaughan, W., Xiao, Zhiyong, Chabot, N., Denevi, B., Ernst, C., Helbert, J., Amore, M., Maturilli, A., Head, J., Solomon, S.: Mercury's hollows: constraints on formation and composition from analysis of geological setting and spectral reflectance, JGR Planets. **118**, 1013–1032 (2013)

Burbine, T., McCoy, T., Nittler, L., Benedix, G., Cloutis, E., Dickinson, T.: Spectra of extremely reduced assemblages: Implications for Mercury. Meteorit. Planet. Sci. **37**, 1245–1254 (2002)

Butler, B.: The migration of volatiles on the surfaces of Mercury and the moon. JGR Planets. **102**(E8), 19283–19291 (1997)

Butler, B., Muhleman, D., Slade, M.: Mercury—full-disk radar images and the detection and stability of ice at the North-Pole. JGR Planets, **98**(E8), 15003–15023 (1993)

Byrne, P., Klimczak, C., Williams, D., Hurwitz, D., Solomon, S., Head, J., Preusker, F., Oberst, J.: An assemblage of lava flow features on Mercury. J. Geogr. Res. Planets. **118**, 1303–1322 (2013)

Chabot, N., Ernst, C., Denevi, B., Harmon, J., Murchie, S., Blewett, D., Solomon, S., Zhong, E.: Areas of permanent shadow in Mercury's south polar region ascertained by MESSENGER orbital imaging. Geophys. Res. Lett. **39**, L09204 (2012). doi:10.1029/2012GL051526

Chabot, N., Ernst, C., Harmon, J., Murchie, S., Solomon, S., Blewett, D., Denevi, B.: Caters hosting radar-bright deposits in Mercury's north polar region: areas of persistent shadow determined form MESSENGER images. J. Geogr. Res. Planets. **118**, 26–36 (2013)

Cintala, M.: Impact-induced thermal effects in the lunar and Mercurian regoliths. J. Geogr. Res. Planets. **97**, 947–973 (1992)

Cintala, M. Grieve, R.: Scaling impact melting and crater dimensions: Implications for the lunar cratering record. Meteorit. Planet. Sci. **33**, 889–912 (1998)

Clark, P.E.: Mercury's Surface. In Dynamic planet: Mercury in the context of its environment, 61–106. Springer, Berlin (2007)

Crider D., and Killen, R.: Burial rate of Mercury's polar volatile deposits. GRL. **32**(12), L12201 (2005)

Correia, A. Laskar, J.: Impact cratering on Mercury: Consequences of the spin evolution. Astrophys. J Lett. 751, L43 (2012)

Denevi, B., Robinson, M.: Mercury's albedo from Mariner 10: Implications for the presence of ferrous iron. Icarus. **197**, 239–246 (2008)

Denevi, B., Robinson, M., Solomon, S., Murchie, S., Blewett, D., Domingue, D., McCoy, T., Ernst, C., Head, J., Watters, T., Chabot, N.: The evolution of Mercury's crust: A global perspective from MESSENGER. Science. **324**, 613–618 (2009)

Denevi, B., Ernst, C., Meyer, H., Robinson, M., Murchie, S., Whitten, J., Head, J., Watters, T., Solomon, S., Ostrach, L., Chapman, C., Byrne, P., Klimczak, C., Peplowski, P.: The distribution and origin of smooth plains on Mercury. J. Geogr. Res. Planets. **118**, 891–907 (2013)

Domingue, D., Vilas, F., Holsclaw, G., Warrell, J., Izenberg, N., Murchie, S., Denevi, B., Blewett, D., McClintock, W., Anderson, B., Sarantos, M.: Whole-disk spectrophotometric properties of Mercury: Synthesis of MESSENGER and ground-based observations. Icarus. **209**, 101–124 (2010)

Domingue, D., Murchie, S., Chabot, N., Denevi, B., Vilas, F.: Mercury's spectrophotometric properties: Update from the Mercury dual imaging system observations during the Third MESSENGER flyby. Planet. Sp. Sci. **59**, 1853–1972 (2011a)

Domingue, D., Murchie, S., Denevi, B., Chabot, N., Blewett, D., Laslo, N., Vaughan, R., Kang, H., Shepard, M.: Photometric correction of Mercury's global color mosaic. Planet. Sp. Sci. **59**, 1873–1887 (2011b)

Domingue, D., Chapman, C., Killen, R., Zurbuchen, T., Gilbert, J., Sarantos, M., Benna, M., Slavin, J., Schriver, D., Travnicek, P., Orlando, T., Sprague, A., Blewett, D., Gillis-Davis, J.,

Feldman, W., Lawrence, D., Ho, G., Ebel, D., Nittler, L., Vilas, F., Pieters, C., Solomon, S., Johnson, C., Winslow, R., Helbert, J., Peplowski, P., Weider, S., Mouawad, N., Izenberg, N., McClintock, W.: Mercury's weather-beaten surface: Understanding Mercury in the context of Lunar and asteroidal space weathering studies. Sp. Sci. Rev. **181**, 121–214 (2014)

Egea-Gonzalez, I., Ruiz, R., Fernandez, C., Williams, J.P., Marquez, A., Lara, L.: Depth of faulting and ancient heat flows in the kuiper region of mercury from lobate scarp topography. Planet. Sp. Sci. **60**, 193–198 (2012)

Elkins-Tanton, L. Hager, B.: Giant impacts can cause volcanism. EPSL. **239**, 219–232 (2005)

Erard, S., Bezard, B., Dorressoundiram, A., Despan, D.: Mercury resolved spectroscopy from NTT. Planet. Sp. Sci. **59**, 1842–1852 (2011)

Fassett, C., Head, J., Blewett, D., Chapman, C., Dickson, J, Murchie, S., Solomon, S., Watters, T.: Caloris impact basin: Exterior geomorphology, stratigraphy, morphometry, radial sculpture, and smooth plains deposits. EPSL. **285**(3-4), 297–308 (2009)

Fassett, C., Kadish, S., Head, J., Solomon, S., Strom, R.: The global population of large craters on Mercury and comparison with the Moon. Geophys. Res. Lett. **38**, L10202 (2011). doi:10.1029/2011GL047294

Fassett, C., Head, J., Baker, D., Zuber, M., Smith, D., Neumann, G., Solomon, S., Klimczak, C., Strom, R., Chapman, C., Prockter, L., Phillips, R., Oberst, J., Preusker, F.: Large Impact Basins on Mercury: Global distribution, characteristics, and modification history from MESSENGER orbital data. J. Geogr. Res. Planets. **117**, EE0l08 (2012). doi:10.1029/2012JE004154

Freed, A., Solomon, S., Watters, T., Phillips, R., Zuber, M.: Could Pantheon Fossae be the result of the Apollodorus crater forming impact within the Caloris basin, Mercury?, EPSL, **285**, 320–327 (2009).

Freed, A., Blair, D., Watters, T., Klimzcak, C., Byrne, P., Solomon, S., Zuber, M., Melosh, H.: On the origin of graben and ridges within and near volcanically buried craters and basins in Mercury's Northern plains. JGR Planets. **117**, E00L06 (2012). doi:10.1029/2012JE004119

Gillis-Davis, J., Blewett, D., Gaskell, R., Denevi, B., Robinson, M., Strom, R., Solomon, S., Sprague, A.: Pit-floor craters on Mercury: Evidence of near-surface igneous activity. EPSL. **285**(3-4), 243–250 (2009)

Hartman, B., Domingue, D.: Scattering of light by individual particles and the implications for models of planetary surfaces. Icarus. **131**, 421–448 (1998)

Head, J., Chapman, C., Strom, R., Fassett, C., Denevi, B., Blewett, D., Ernst, C., Watters, T., Solomon, S., Murchie, S., Prockter, L., Chabot, N., Gillis-Davis, J., Whitten, J., Goudge, T., Baker, D., Hurwitz, D., Ostrach, L., Xiao, Z., Merline, W., Kerber, L., Dickson, J., Oberst, J., Byrne, P., Klimczak, C., Nittler, L.: Flood volcanism in the northern high latitudes of Mercury revealed by MESSENGER. Science. **333**, 1853–1856 (2011)

Helfenstein, P., Veverka, J.: Photometric properties of lunar terrains derived from Hapke's equation. Icarus. **72**, 342–357 (1987)

Herrick, R., Curren, L., Baer, A.: A Mariner/MESSENGER global catalog of Mercurian craters. Icarus. **215**, 452–454 (2011)

Holsclaw, G., McClintock, W., Domingue, D., Izenberg, N., Blewett, D., Sprague, A.: A comparison of the ultraviolet to near-infrared spectral properties of Mercury and the Moon as observed by MESSENGER. Icarus. **209**(1), 179–194 (2010)

Hurwitz, D., Head, J., Byrne, P., Xiao, Z., Solomon, S., Zuber, M., Smith, D., Neumann, G.: Investigating the origin of candidate lava channels on Mercury with MESSENGER data: Theory and observations. J.Geogr. Res. Planets. **118**, 471–486 (2013)

Kerber, L., Head, J., Solomon, S., Murchie. S., Blewett, D., Wilson, L.: Explosive volcanic eruptions on Mercury: Eruption conditions, magma volatile content, and implications for interior volatiles abundances. EPSL. **285**, 263–271 (2009)

Kerber, L., Head, J., Blewett, D., Solomon, S., Wilson, L., Murchie, S., Robinson, M., Denevi, B., Domingue, D.: The global distribution of pyroclastic deposits on Mercury: The view from MESSENGER flybys 1–3. Planet. Sp. Sci. **59**, 185–1909 (2011)

Klimczak, C., Schultz, R., Nahm, A.: Evaluation of the origin hypotheses of Pantheon Fossae, central Caloris Basin, Mercury. Icarus. **209**, 262–270 (2010)

Klimczak, C., Ernst, C., Byrne, P., Solomon, S., Watters, T., Murchie, S., Preusker, F., Balcenski, J.: Insights into the subsurface structure of the Caloris basin, Mercury, from assessments of mechanical layering and changes in long-wavelength topography. J. Geogr. Res. Planets. **118**, 2030–2044 (2013)

Kring, D.: Parameters of Lunar Soils, Lunar Exploration Initiative. http://www.lpi.usra.edu/science/kring/lunar_exploration/briefings/lunar_soil_physical_properties.pdf (2006). Accessed 14 Nov 2014

Langevin, Y.: The regolith of Mercury: present knowledge and implications for the Mercury orbiter mission. Planet. Sp. Sci. **45**(1), 31–37 (1997)

Lawrence, D., Feldman, W., Goldsten, J., Maurice, S., Peplowski, P., Anderson, B., Bazell, D., McNutt, R., Nittler, L., Prettyman, T., Rodgers, D., Solomon, S., Weider, S.: Evidence for water ice near Mercury's north pole from MESSENGER neutron spectrometer measurements. Science. **339**, 292–296 (2013)

Loeffler, M., Dukes, C., Baragiola: Irradiation of olivine by 4 keV He+: Simulation of space weathering by the solar wind. JGR Planets. **114**, E03003. doi:10.1029/2008JE003249 (2009)

Marchi, S., Massironi, M., Cremonese, G., Martellato, E., Giacomini, L., Prockter, L.: The effects of the target material properties and layering on the crater chronology: The Case of Raditlkadi and Rachmaninoff Basins on Mercury. Planet. Sp. Sci. **59**, 1968–1980 (2011)

Melosh, J.: Impact cratering: A geologic process, p. 253. Oxford University Press, London (1989)

Mueller, J., Simon, S., Wang, Y., Motschmann, U., Heyner, D., Schuele, J., Ip, W., Kleindienst, G., Pringle, G.: Origin of Mercury's double magnetopause: 3D hybrid simulation study with A.I.K.E.F., Icarus, **218**, 666–687 (2012)

Neish, C., Blewett, D., Harmon, J., Coman, E., Cahill, J., Ernst, C.: A comparison of rayed craters on the Moon and Mercury. J. Geogr. Res. Planets. **118**, 2247–2261 (2013)

Neumann, G., Cavanaugh, J., Sun, X., Mazarico, E., Smith, D., Zuber, M., Mao, D., Paige, D., Solomon, S., Ernst, C., Barnouin, O.: Bright and dark polar deposits on the Mercury: Evidence for surface volatiles. Science. **339**, 296–300 (2013)

Nicholis, M., Rutherford, M.: Pressure dependence of graphite-C-O phase equilibria and its role in lunar volcanism. Lun. Planet. Inst. **36**, 1726.pdf (2005)

Noble, S., Pieters, C., Keller, L.: An experimental approach to understanding the optical effects of space weathering. Icarus. **192**, 629–642 (2007)

Paige, D., Siegler, M., Harmon, J., Neumann, G., Mazarico, E., Smith, D., Zuber, M., Harju, E., Delitsky, M., Solomon, S.: Thermal stability of volatiles in the north polar region of Mercury. Science. **339**, 300–303 (2013)

Pieters, C., Tompkins, S.: Tsiolkovsky crater: A window into crustal processes on the lunar farside. JGR Planets. **104**(E9), 21935–21949 (1999)

Pike, J.: Control of crater morphology by gravity and target type-Mars, Earth, Moon. Proc. Lun. Plan. Sci. Conf. **10**, 2159–2189 (1980)

Pike, J.: Geomorphology of impact craters on Mercury. In Vilas, F., Chapman, C., Matthews, M. (eds.) Mercury, pp. 165–273. U Arizona Press, Tucson (1988)

Riner, M., Lucey, P.: Spectral effects of space weathering on Mercury: The role of composition and environment. Geophys. Res. Lett. **39**, L12201 (2012). doi:1029/2012GL052065

Riner, M., Lucey, P., McCubbin, F., Taylor, G.: Constraints on Mercury's surface composition from MESSENGER neutron spectrometer data. EPSL. **308**, 107–114 (2011)

Ruiz, J., Lopez, V., Dohn, J., Fernandez, C.: Structural control of scarps in the rembrandt region of Mercury. Icarus. **219**, 511–514 (2012)

Rutherford, M., Papale, P.: Origin of basalt fire-fountain eruptions on Earth versus the Moon. Geology. **37**, 219–222 (2009)

Schenk, P.: Thickness constraints on the icy shells of the Galilean satellites from a comparison of crater shapes. Nature. **417**(6887), 419–421 (2002)

Schon, S., Head, J., Baker, D., Ernst, C., Prockter, L., Murchie, S., Solomon, S.: Eminescu impact structure: insight into the transition form complex crater to peak ring basin on Mercury. Planet. Sp. Sci. **59**, 1949–1959 (2011)

Slade, M., Butler, B., Muhleman, D: Mercury radar imaging-Evidence for polar ice, Science. **258**(5082), 635–640 (1992)

Slavin, J., Lepping, R., Wu, C., Anderson, B., Baker, D., Benna, M., Boardsen, S., Killen, R., Korth, H., Krimigis, S., McClintock, W., McNutt, R., Sarantos, M., Schriver, D., Solomon, S., Travnicek, P., Zurbuchen, T.: MESSENGER observations of large flux transfer events at Mercury. Geophys. Res. Lett. **37**, L02015 (2010). doi:10.1029/2009GL041485

Smith, D., Zuber, M., Phillips, R., Solomon, S., Hauck, S., Lemoine, F., Mazarico, E., Neumann, G., Peale, S., Margot, J.-L., Johnson, C., Torrence, M., Perry, M., Rowlands, D., Goossens, S., Head, J., Taylor, A.: Gravity field and internal structure of Mercury from MESSENGER. Science. **336**, 214-217 (2012)

Spudis, P. Guest, J.: Stratigraphy and geologic history of Mercury. In Vilas, F., Chapman, C., Matthews, M. (eds.) Mercury, pp. 118-164. U Arizona Press, Tucson (1988)

Starukhina, L.: Water detection on atmosphereless celestial bodies: alternative explanations of the observations. JGR Planets. **106**(E7), 14701–14710 (2001)

Strom, R., Banks, M., Chapman, C., Fassett, C., Forde, J., Head, J., Merline, W., Prockter, L., Solomon, S.: Mercury crater statistics from MESSENGER flybys: Implications for stratigraphy and resurfacing history. Planet. Sp. Sci. **59**, 1960–1967 (2011)

Sun, X., Cavanaugh, J., Neumann, G., Smith, D., Zuber, M.: Mapping the topography of Mercury with MESSENGER laser altimetry. SPIE Newsroom. (2012). doi:10.1117/2.1201210.004489

Vasavada, A., Paige, D., Wood, S.: Near-surface temperatures on Mercury and the Moon and the stability of polar ice deposits. Icarus. **141**(2), 179–193 (1999)

Watters, T., Solomon, S., Robinson, M., Head, J., Andre, S., Hauck, S., Murchie, S.: The tectonics of Mercury: the view after MESSENGER's first flyby. EPSL. **285**, 283–296 (2009)

Wieczorek, M., Correia, A., LeFeuvre, M., Laskar, J., Rambaux, N.: Mercury's spin-orbit resonance explained by initial retrograde and subsequent synchronous rotation. Nat. Geosci. **5**, 18–21 (2012)

Xiao, Z., Strom, R., Blewett, D., Byrne, P., Solomon, S., Murchie, S., Sprague, A., Domingue, D., Helbert, J.: Dark spots on Mercury: A distinctive low-reflectance material and its relation to hollows. J. Geogr. Res. Planets. **118**, 1752–1754 (2013)

Yan, N., Chassefiere, E., Leblanc, F., Sarkissian, A.: Thermal model of Mercury's surface and subsurface: impact of subsurface physical heterogeneities on the surface temperature. Adv. Sp. Res. **38**, 583–588 (2006)

Chapter 4
Mercury's Surrounding Environment

4.1 Mercury's Surface-Bounded Exosphere

MESSENGER has provided further details on Mercury's surface-bounded atmosphere, a system dynamically coupled with the surface and magnetosphere, with the surface acting as both source and sink, and the magnetosphere acting as both accelerator and buffer for the surrounding plasma (Killen et al. 2010). Interactions between the magnetosphere and IMF cause changes in the exosphere on the scale of minutes. Table 4.1 provides a comparison between composition and pressure in Mercury's exosphere, the true atmospheres of the other terrestrial planets, and the lunar exosphere. Note similarities between the two exospheres. The exosphere has two populations: one source directly from the regolith, the other ambient, periodically adsorbed by the regolith while undergoing one or more ballistic 'hops' (Killen et al. 2010). The production mechanisms are species-dependent. Exosphere/surface interaction processes are illustrated (Fig. 4.1. Table 4.2 show the relationship between process, volatility, and energy level. Note how the volatility of species increases as the energy level of the removal
process increases..

4.2 Earlier Observations and Models of Sodium and Potassium Exosphere

Mercury's sodium (Na) D1 and D2 lines (589.0 and 589.6 nm) were first observed by Potter and Morgan in 1985. Ten years later, Smyth and Marconi (1995) published a model that accurately predicted, to first approximation, the atomic spatial distribution and relative importance of the initial solar atmosphere and ambient atmosphere for Na and potassium (K) as a function of position in its solar orbit (True Anomaly Angle or TAA). They inferred that variation in the high-energy tail of the exosphere was controlled by the extremely variable solar radiation acceleration, as both species resonantly scatter solar photons. The extent of the lateral, anti-sunward

P. E. Clark, *Mercury's Interior, Surface, and Surrounding Environment*,
SpringerBriefs in Astronomy, DOI 10.1007/978-1-4939-2244-4_4

Table 4.1 Comparison of terrestrial planet atmospheres

Terrestrial planet	Major constituents	Pressure at surface
Mercury exosphere	Major H, He, O_2, Na, K, Ar	Trace
	Minor/trace H_2, O_2, H_2O, OH, N_2+, CO_2+, S, Mg, Ca, Fe, Si, Al	
Venus atmosphere	Major CO_2, N_2	92 Bars
	Minor/trace SO_2, Ar, H_2O, CO, He, rare gases	
Earth atmosphere	Major N_2, O_2, Ar	1 Bar
	Minor/trace CO_2, H_2O, SO_2, CO, CH_4	
Mars atmosphere	Major CO_2, N_2, Ar, O_2	0.07 Bar
	Minor/trace Co, H_2O, rare gases	
Lunar exosphere	Major H, He, Na, K, Ar, Rn	Trace
	Minor/trace H_2, H_2O	

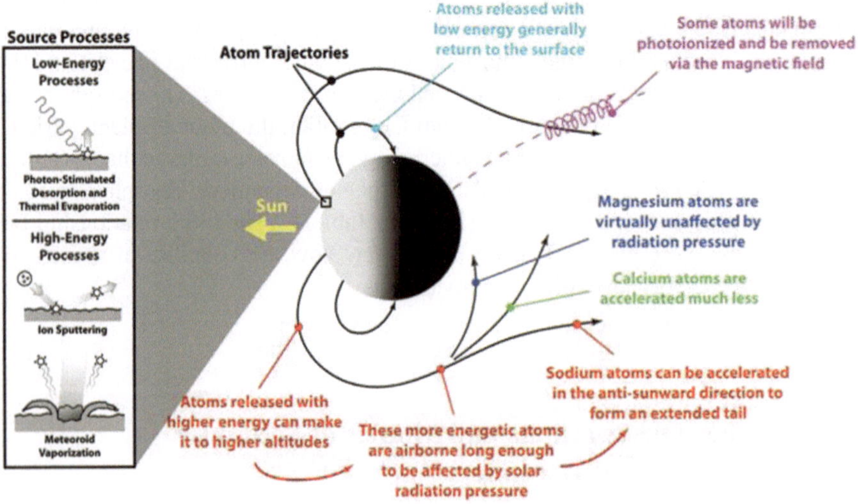

Fig. 4.1 Mercury exosphere/surface source/sink processes. Courtesy of NASA and Johns Hopkins University Applied Physics Laboratory.

distribution and transport rate of ambient atoms was apparently driven by solar radiation acceleration, forming a tail. The high eccentricity of Mercury's orbit means that radial velocity, and radiation acceleration, can vary considerably during each orbital period. The origin of the sodium exosphere remained controversial (Hunten et al. 1988; Killen and Ip 1999; Leblanc et al. 2007) until systematic and statistically significant observations were acquired by Potter and coworkers (2006, 2007,

Table 4.2 Progressively more volatile species, sources, and mechanisms

Species	Source	Mechanisms	Distribution
H, He, Noble gases (Ar)	Solar wind, diffusion from interior, radiogenic release, regolith	Source: thermal release Sink: implantation (H), space weathering (H)	T-dependent, altitude and longitude dependent
H_2O, S	Surface, meteoritic infall, cold traps in regolith, diffusion from interior	Source: thermal release, electron or photon stimulated desorption (ESD or PSD) Sink: photo-ionization	T-dependent as above
Na, K, alkalis	Regolith, radiogenic release (K only)	Source: thermal release (K only), PSD, photo- and ion-sputtering, impact vaporization (IV) Sink: photoionization	Mixture: thermal release and PDS T-dependent, sputtering from energetic particles (cusps at higher latitudes, dusk side due to dawn-dusk electric field direction, or energetic event), IV spatially distributed
Ca, Mg, Al, refractories	Regolith	Ion sputtering, IV Sink: photoionization	Sputtering as above. IV as above

2008, 2009). The models were confirmed by ground-based and current observations described below (e.g., Potter et al. 2009; Burger et al. 2012, 2014; Killen et al. 2010, in press; LeBlanc and Johnson 2010; Wang and Ip 2011; Mouawad et al. 2011; Mura 2012; Schmidt et al. 2012, 2013).

Ground-based observations of Mercury's exosphere were done in support of the MESSENGER mission prior to orbital insertion. Using their capability to detect Doppler shifts as small as 0.1 km/s, Potter and coworkers (2009) confirmed earlier observations of solar radiation acceleration-induced anti-sunward velocities for sodium atoms. They observed an E/W flow pattern as sodium flowed outward from the subsolar point except above the dawn terminator where velocity vectors were Earthward, which they interpreted as the evaporation of sodium from heating the cold surface at daybreak. An additional N/S flow pattern indicated sodium flowing outward from the subsolar point. High velocities and emission were always observed to be correlated and could appear in either hemisphere, indicating high latitude sources of sodium in both hemispheres. Ip (1990) had already predicted that at peak radiation acceleration, 3–5 times more sodium would accumulate near the terminator than at the subsolar point. Emission intensity appears to be affected by solar radiation acceleration more than anticipated on the basis of planetary radial velocity alone.

Killen and coworkers (2010) continued to characterize the relationship between sodium and potassium. The highly variable exospheric Na/K ratio averages

100+/−20, much higher than elsewhere in the Solar System. The lunar exospheric Na/K ratio is comparable to the crustal ratio: 5 to 10. Both sodium and potassium are seen in the anti-sunward tail via acceleration by radiation pressure due to strong resonance lines. The diffusion rate for K is far less than the rate for Na, limited by the surface diffusion rate, which is a function of temperature and grain size. Thus, Mercury's Na/K ratio is weakly correlated with the solar UV flux, a source of variability that exceeds variations in local abundance. The distribution, and thus source processes, for Na and K are distinctively different from each other and from other species. The high temperature (>10,000 K) source process for potassium observed in the surrounding exosphere and tail is consistent with ion sputtering and/or impact vaporization with subsequent dissociation. According to Killen and coworkers (2010), Na and K could be derived from a regolith with a composition similar to *Apollo 17* orange glass.

In the following sections we will be discussing results from MESSENGER, typically involving the MASCS instrument, which provided three Mercury flybys prior to orbital insertion. Flybys are particularly well suited to providing snapshots of Mercury's exosphere and its interaction with surface and magnetosphere.

4.3 MESSENGER Observations and Models of Sodium Exosphere

Based on comparison of MASCS observations to their theoretical model, Burger and coworkers (2010) determined that the dominant cause of sodium in the exosphere was photon stimulated desorption (PSD), with the rate limited by the slower diffusion rate of sodium from interior grains to the topmost monolayers, where PSD is effective. Solar wind changes were variable on the time scale of the observations, and thus observations were highly variable. The PSD process, dependent on UV flux and sodium concentration, occurs where the surface is open to the solar wind (dayside) or where the magnetospheric plasma impacts the surface (nightside), peaking at high latitudes in both hemispheres and dominating the tail and fantail. As the radiation pressure increases, sodium ions are pushed into the tail. Ion-enhanced diffusion is thought to increase the source rate at higher latitudes. The diffusion rate must be at least a factor of 5 higher in ion precipitation in these 'cusp' regions to match MESSENGER MASCS observations. Differences between the PSD model and observations may also be due to partial thermal accommodation and reemission of returning sodium ions as well as additional energy sources. The sodium diffusion rate is temperature dependent, resulting in a slight increase in exospheric density on the dayside. Sputtering by the solar wind at the poles could explain the poleward peak in emission, but not the high intensities observed. The overall trend in distribution is not consistent with impact vaporization, which would be more symmetrical above and below the equator, but a 15% contribution from impact vaporization due to micrometeoroid bombardment could explain the disparity. Loss mechanisms include sticking to the surface or escaping into the magnetotail. The source rate could range from 10^6 to $10^7 cm^{-2}s^{-1}$ depending on the sticking efficiency of sodium returning to the surface.

Leblanc and Johnson (2010) noted that the annual cycle of emission brightness for the sodium exosphere is roughly the same from year to year. They considered a number of ejection mechanisms, concluding that no one mechanism dominated throughout an annual cycle. They proposed that the main process was PSD, most efficiently inducing dawn/dusk asymmetries. Wang and Ip (2011) preferred ion-enhanced PSD (IE-PSD) as a sodium source generating mechanism and observed that the sodium tail production rate and terminator to limb ratio could be explained with no temperature-enhanced 'sticking' effect and a small thermal accommodation. According to LeBlanc and Johnson (2010), thermal desorption dominates at perihelion, inducing peak emission between 25 and 40°. Solar wind-induced sputtering would induce the large dispersion of emission brightness occurring on very short time scales. Micrometeoroid vaporization has the least obvious signature and uniformly increases the surface reservoir. Their study highlighted the possible effect of ion bombardment and surface charging in enhancing diffusion of the surface reservoir. Their observations were consistent with PSD as the main source of sodium ions, increased by a factor of up to 2.5 due to enhanced diffusion or negative surface charging. In their model, the solar wind impacts the surface, rapidly depleting it of source ions—unless particle bombardment and surface charging induce diffusion and change in the energy distribution of reabsorbed particles, a competing process that maintains emissions for hours as observed. The sodium abundances in the exosphere vary as a function of orbital position, due to variation in nightside migration. Solar radiation pressure induces an overall global circulation of sodium from dayside to nightside, the rate of transfer varying 'seasonally' as a function of Mercury's position in its orbit, with the rate increasing as aphelion is approached.

Mouawad and coworkers provided constraints on the sodium exosphere (2011) based on MESSENGER MASCS and ground-based observations in general agreement with LeBlanc and Johnson (2010). Based on the derived bound exosphere temperature (900–1200 K) and resulting velocity distributions, which are influenced by the degree of thermal accommodation, PSD was considered the most likely source. The degree of thermal accommodation provides the upper limit on the contribution of impact vaporization. Researchers referred to slow (Maxwellian distribution) PSD, involving a gentle release of sodium atoms and less accommodation, and fast (Weibull distribution) PSD, involving quicker release and more accommodation. Slow PSD requires less accommodation, because as accommodation increases, little sodium can escape into the tail and the source rate must be increased, which would create too much emission on the dayside. Fast PSD on the other hand requires more thermal accommodation and more sticking to prevent energetic atoms from overpopulating the tail. The lack of spatial resolution results in the inability to distinguish between PSD and IE-PSD, and thus to determine the importance of ion precipitation on the sodium diffusion rate (Fig. 4.2). The best models predict either a Weibull distribution with both PSD and IE-PSD components, or a Maxwellian distribution with a PSD component only. Greater thermal accommodation results in a Weibull (fast) velocity distribution, and a moderate accommodation results in a Maxwellian (slow)

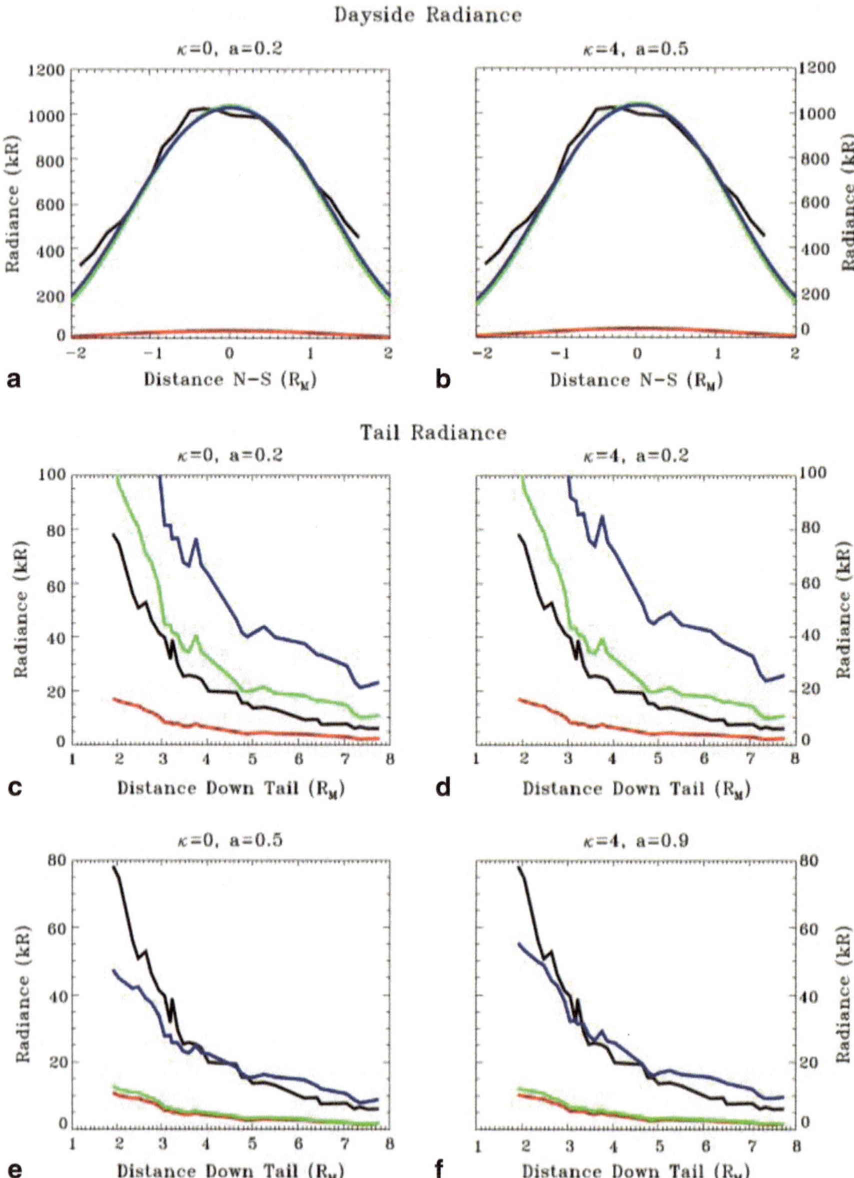

Fig. 4.2 Observations for dayside and tail radiance (*black lines*) fit to models: *red* impact vaporization (IV), *green* IV and slow PSD, *blue* IV and fast PSD with different values of k (diffusive flux normalized to the peak ion flux) and a (thermal accommodation coefficient). (Reprinted from Icarus, 211, Mouawad et al. 2011, Fig. 7, © 2011 with permission from Elsevier)

velocity distribution. A greater accommodation and fast PSD implies a minimal impact vaporization contribution, whereas a weaker accommodation and slow PSD implies a greater impact vaporization (IV) contribution. Killen and Sarantos (2004) indicated that the IV component, if it indeed scaled with orbital distance, could be ignored, as it would contribute on the order of a couple of percent.

Mura (2012) explored the times scales for production and loss of the sodium exosphere. Sodium in the regolith is continuously lost through gravitational escape, incomplete escape, radiation pressure acceleration or photoionization and replaced by diffusion from the interior of regolith grains or chemical sputtering. Returning particles may either bounce or stick. Mura uses Monte Carlo simulations to predict exospheric densities as a function of loss rates at different orbital positions (true anomaly angles). The model predicts time scales over which variations and restoration to a steady state occur in response to impulsive events. Mura (2012) found that time scales vary from 1–2 h (close to perihelion) to half a day (close to aphelion). The probability of complete escape ranges from 20 % at perihelion and aphelion to 40 % at true anomaly angles of 60 and 300°. Mercury's sodium tail is the most notable feature of its exosphere. Mura (2012) modeled sodium's escape probability as a function of true anomaly angle (TAA) and ejection velocity distribution functions (Fig. 4.3), providing total sodium abundance as a product of its flux multiplied by its photoionization lifetime. Although the sodium exosphere abundances vary as a function of TAA, the source rate would be constant if a threshold were reached, as in diffusion or chemical sputtering limited scenarios. Up to 20 % of the sodium atoms are deposited on the nightside surface after one or more ballistic orbits, and then reemitted via PSD at the dawn terminator representing an additional source varying in strength as a function of angular velocity. Thus both loss and replenishment rates and the times scales over which these occur vary as a function of TAA. LeBlanc and Doressoundiram (2010) report that the Na/K abundance ratios vary in a similar manner. Because such variations can occur on time scales of hours, Mura (2012) recommends that time resolution for atmospheric measurements be <1 h.

Schmidt and coworkers (2012) models, consistent with sodium abundances and variations as a function of orbital phase, indicated that PSD combined with micrometeoroid impacts, with Maxwellian velocity distributions at 1500 and 3000 K, were source mechanisms for sodium and most responsible for the sodium tail. Their work indicated that up to 20 % of the highest energy component can escape down the tail as a result of increased solar radiation pressure. To maintain a constant supply of sodium in the exosphere, the average column density needs to decrease by 1/3 at the peak escape position in the orbit (70° TAA). They found that desorption of sodium into the exosphere by electrons is negligible compared to protons, but the energy distribution is similar. Space weather-induced variability, resulting from the pushing of the magnetosphere to the surface of a large meteor impact, can be short-term and extreme and, based on the observed brightness, result in an order of magnitude increase in the sputtering rate. Impacts, on the other hand, are difficult to detect above dayside background emission (Mangano et al. 2007). For a large impact, a plume of intensely emitting sodium would move downtail. As impacts appear to occur with slightly greater frequency in the morning (Marchi et al. 2005), gravity

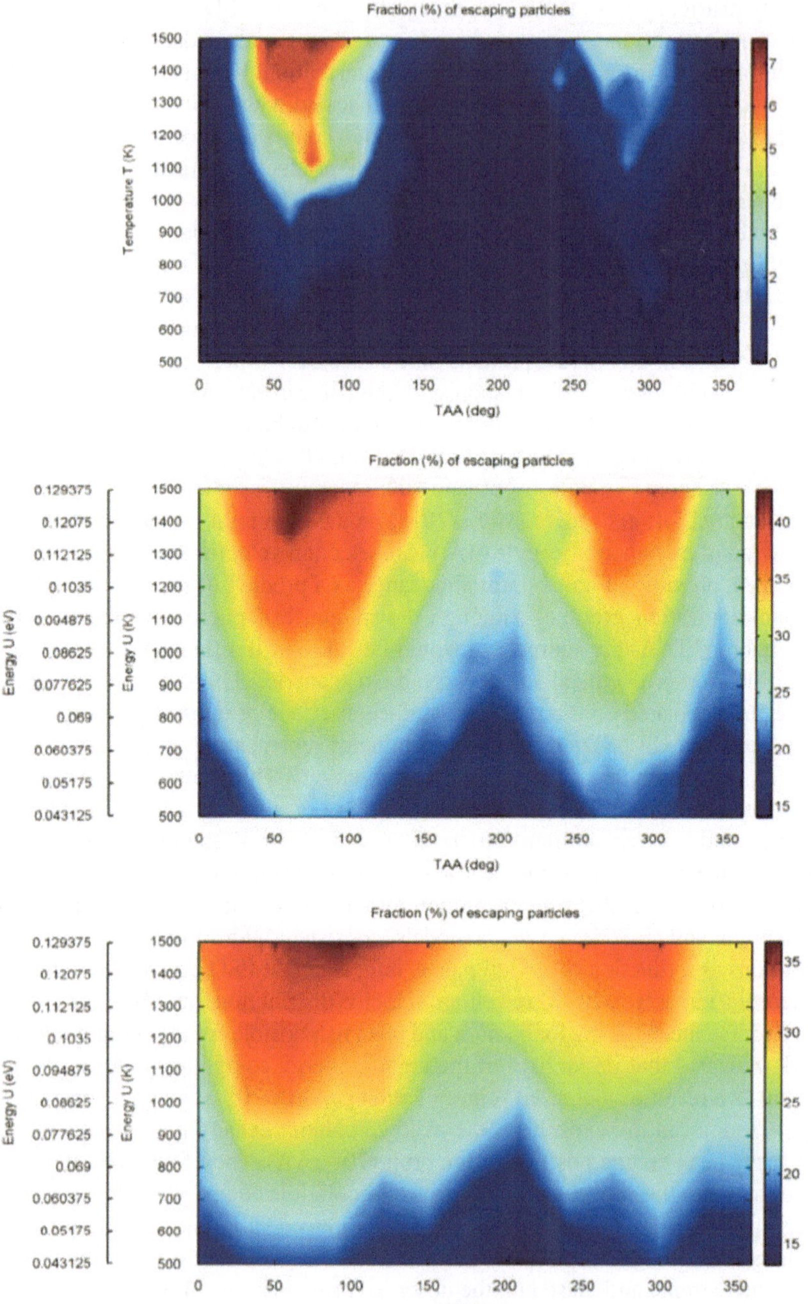

Fig. 4.3 Loss rate of sodium due to PSD release as function of TAA for (*top*) Maxwellian temperature distribution, (*middle*) non-Maxwellian temperature distribution, (*bottom*) constant temperature. (Reprinted from planetary and space science, 63–64, Mura et al. 2012, Fig. 1, © 2012 with permission of Elsevier)

would cause the plume to move across the nightside and make its way downtail. Based on LCROSS results, Schmidt and coworkers (2012) suggest that a small fraction of sodium liberated during an impact would form oxides and hydroxides with short lifetimes due to photodissociation, producing a small population of energetic neutralized sodium. However, as in the case of other studies, the sodium distribution produced by combined PSD and impact vaporization lacked the high-energy component necessary to explain the brightness and variability of the sodium neutral tail. Additional sputtering during energetic solar events is still a factor of 4 too low to explain the tail brightness. Schmidt and coworkers proposed an additional thermalized, globally distributed energetic impact source (at 5000 K) comparable to PSD, resulting from variable scale impacts on longer timescales, such as meteor showers from sun-grazing comets, to explain overall tail brightening, rather than large impacts resulting in distinct hot spots.

4.4 Neutral and Charged Exospheric Distributions: Calcium

MESSENGER MASCS observations have led to the characterization of species other than Na and K, including the magnesium (Mg), calcium (Ca), and Ca^+ originating as regolith constituents. Their distributions are affected by their atomic properties, and thus, energetic mechanisms are required to remove them, as well as their abundances in the regolith, which acts as a primary source. Exospheric sources of these species include impact vaporization generated high temperature metallic oxides and hydroxides of magnesium or calcium, which dissociate to produce atomic Mg, Ca, or O.

Neutrals are subject to losses (Sarantos et al. 2011) such as ballistic escape; radiation pressure acceleration, which ranges from strong for Na to weak for Ca to almost non-existent for Mg; surface impact induced adsorption; and photoionization. Mg has a very short photoionization lifetime. Neutrals are continuously generated from the surface by PSD, thermal desorption, impact vaporization, solar wind sputtering, and interaction of recycled magnetospheric ions.

Regolith magnesium and calcium abundances are relatively high, so despite their refractory natures, their discovery as exospheric species in Mercury's energetic environment is not surprising (e.g., Killen et al. 2010; Vervack et al. 2010; McClintock et al. 2009). As expected, source and sink processes for magnesium and calcium differ from those of the more volatile Na and K. MESSENGER revealed different distributions of magnesium, calcium, and sodium in the anti-sunward tail, dawn-dusk fantail, and terminators, implying different source, transfer, and loss processes (McClintock et al. 2009) (Table 4.2). Tail observations are taken by scanning up and down along the Sun-Mercury line. Fantail observations are taken by pointing toward the dawn terminator while in the equatorial plain, and executing a 180° roll through north to the dusk terminator. Ca and Mg emissions are statistically significant within 8 Mercurian radii, but much weaker than sodium emissions. Calcium has dawn emission ten times higher than the dusk emission, due to the enhanced meteoroid impact velocity resulting from Mercury's orbital motion. Magnesium emission is more uniform but with evidence of a weak double-lobed N/S enhancement, whereas calcium emission has an equatorial peak and a sodium high latitude enhancement. Ca peaks near the dawn equator, decreasing in the dusk direction; whereas sodium emission is weaker toward dawn, increases toward the pole, and decreases again toward dusk. Mg fantail observations, in contrast, are statistically uniform. Na, Ca, and Mg emission increased as the spacecraft approached the dayside and the proportion of the column illuminated increased, but sodium emission increased disproportionately. McClintock and coworkers (2009) speculated that ion sputtering and impact vaporization, the most energetic processes, acting near the dawn terminator provide most of the tail material. Alternatively, exospheric species return to the nightside and are subsequently released by photon desorption at sunrise. The distribution of material, once released, is determined by gravity, radiation pressure, and photoionization, and secondarily by magnetic and electrical field transport. These influences, when combined, result in rapid loss through return to the surface or along open magnetic field lines. Such enhancements are not always present, but apparently correlate with solar wind and plasma impingement on the surface. Calcium ionization lifetime is relatively small, much less subject to radiation pressure, and thus must be derived from extremely energetic processes to travel down the tail. Magnesium atoms are even smaller than calcium atoms, and thus have longer ionization lifetimes and are less subject to radiation pressure. Sodium and magnesium exospheric abundances are comparable, but calcium is 35–40 times less abundant. A difference of this magnitude between Mg and Ca was not anticipated (McClintock et al. 2009).

Two groups (Burger et al. 2014; Killen and Hahn, in press have focused on analysis and modeling of MASCS orbital observations. They continue to observe a high temperature, persistent, and seasonally varying calcium source, consisting of calcium ejected at greater than escape velocities, in the dawn equatorial region (Fig. 4.4), typically taking scans when the spacecraft is between 6000 and 12,000 km from Mercury. The source region varies little in size and energy distribution. Interestingly, observations and models show strong dawn enhancement in impactor flux for Earth (Janches et al. 2006; Pifko et al. 2013).

According to Burger and coworkers (2014), only 1–2% of the calcium entering the exosphere returns to the surface. Calcium source rates range from 4×10^{23}/s to 0.4 10^{23}/s at TAAs of 20 and 195°, respectively, confirming earlier ground-based observations (Killen et al. 2005; Bida et al. 2000). Burger and coworkers estimated that content of calcium varies from 7 to 36 kg per Mercury year, a factor of 5, whereas the source rate varies by a factor of 10, because calcium's photoionization lifetime is greatest near aphelion, where the source rate is lowest and residence time longest.

Both groups continue to support micrometeorite impact vaporization, which is energetic enough to vaporize calcium-bearing molecules, followed by dissociation of these molecules to release energetic Ca atoms as the release mechanism (Berezhnoy and Klumov 2008; Berezhnoy 2013), also considered a viable model in earlier work (e.g., Cintala 1992; Borin et al. 2009; Grotheer and Livi 2014). Between 2 and 9% of calcium vaporized in Ca-bearing molecules must be photodissociated to become atomic Ca to fit the calcium source models of Killen and Hahn (in press), which is

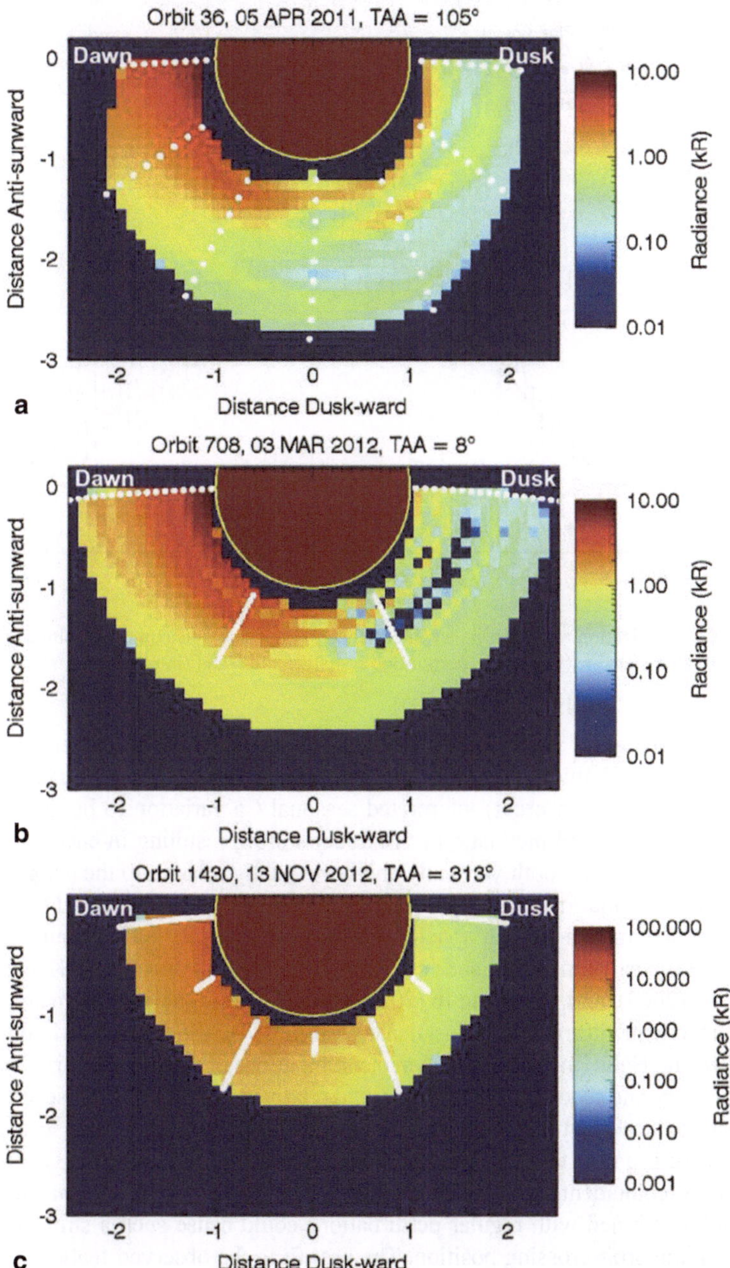

Fig. 4.4 Mercury dayside Ca emission as a function of TAA. Color scale interpolated radiance. Dawn terminator *left*, subsolar point *bottom*. Lines of sight *white* lines. (Reprinted from Icarus, 238, Burger et al. 2014, Fig. 3, © 2014 with permission of Elsevier)

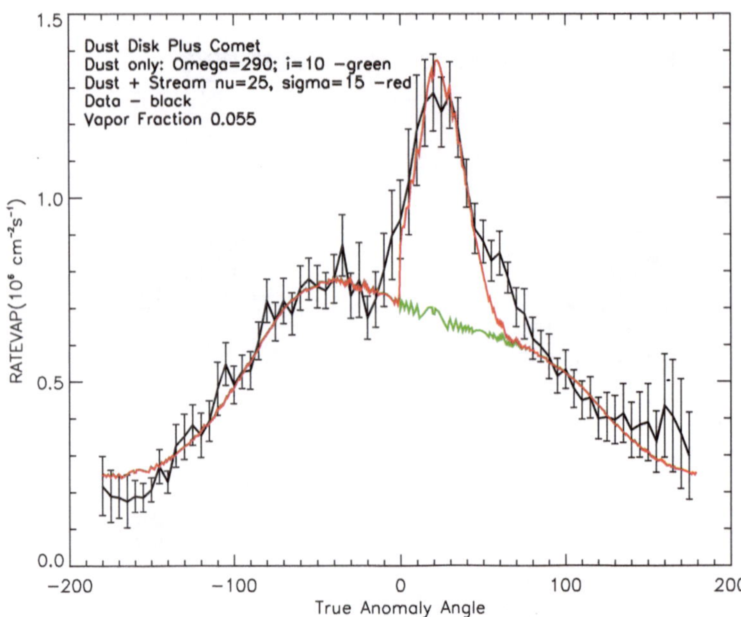

Fig. 4.5 Ca vaporization rate for primary interplanetary dust disk plus cometary source (*spike*) in units of 10^6 atoms/cm^2-s. (Used with permission of Killen and Hahn (in press) Fig. 7)

consistent with earlier work (Berezhnoy et al. 2011). Their predicted zenith column abundances agree within a factor of 2 with estimates by Burger and coworkers (2014).

Killen and Hahn (in press) interpreted seasonal Ca variation to be the result of sizable eccentricity and inclination of Mercury's orbit, resulting in encounters with dust varying in density both vertically and horizontally relative to the ecliptic. They attempted to use that dust density variation to explain the overall trend (maximum near perihelion, minimum near aphelion) along with an observed maximum spike in the exospheric calcium at 45° and local minimum at perihelion −20°. Kameda and coworkers (2009), and Wang and Ip (2011) noted that exospheric sodium abundance is correlated with interplanetary dust density varying over the course of Mercury's orbit. Killen and Hahn (in press) considered combinations of a source localized near the spike, primary and secondary dust disks tilted relative to the ecliptic. Observations of the inner zodiacal light constrains dust disk tilt to <5°. The most likely local source for the spike is a dust tail from comet Encke (Killen and Hahn, in press). Although not entirely coincident in terms of crossing Mercury's orbit at present, the width of the tail combined with regular perturbations could cause such a shift of tens of degrees in the orbit-crossing position. The best fit to the observed features includes the comet tail combined with a primary thin, flat dust disk decreasing little in density with radial distance from the Sun and with no more than a few degrees offset from the ecliptic (Kameda et al. 2009) (Fig. 4.5). The minimum at perihelion −20° is thought to result from enhancement of the nightside reservoir created by the planet's retrograde motion at that point in its orbit, resulting in a 'night' of longer duration.

Burger and coworkers (2012) focused on understanding neutral and ionized calcium. The calcium dawn enhanced emission suggests an additional 'hot source' for calcium. The high temperature of calcium requires an energetic source such as ion sputtering and electron stimulated desorption (ESD), both of which could eject atomic and molecular calcium from the surface at high speeds. Atomic calcium could be produced from dissociation of molecular Ca or CaX (CaO, CaS, etc.) to produce Ca^+. The correlation between S and Ca surface abundances suggest the presence of CaS in the regolith and thus in the exosphere. Implantation and neutralization of Na^+ on the nightside followed by desorption by sunlight at dawn has been proposed to explain Na and Na^+ distributions. Although the process wouldn't be as efficient for calcium, which has a higher sticking coefficient, Ca+ precipitating at dusk onto the surface could supply Ca to the nightside. If this assertion were correct, a seasonal variation would be anticipated, as the terminator actually reverses, allowing more nightside accumulation, at certain times of year. Burger and coworkers (2012) did not observe this seasonal effect.

4.5 Neutral and Charged Exospheric Distributions: Magnesium

Proposed sources of exospheric Mg (e.g., Killen et al. 2010; Sarantos et al. 2011) include solar wind ion sputtering, particularly in polar regions only partially shielded by the magnetic field; micrometeoroid-induced impact vaporization, and dissociation of Mg-bearing molecules such as MgO. Near dawn terminator emissions suggest an additional source, either more local or colder than sputtering or impact. Unlike sodium, regolith Mg abundances are high enough so that normal regolith processes can replenish exospheric magnesium. Mg as a silicate phase constituent requires more energetic processes to release it, whereas Mg that has returned to the surface to be readsorbed can be released by lower energy processes. Sarantos and coworkers (2011) raise the issue of a gardening rate-limited delivery of Mg to the exosphere, but can't resolve it due to the uncertainty of the lifetime of Mg-bearing species in the exosphere.

A number of authors (Killen et al. 2010; Sarantos et al. 2011) have found that a single source production model does not explain the bimodal magnesium distribution (Fig. 4.6). Because of the difference in volatility, the temperature of Mg species is tens of thousands of K, rather than 1000–1500 K for Na and K. Magnesium has no strong resonance lines, as do the two volatile species, to make it susceptible to acceleration by radiation pressure into the exospheric tail, and yet it is observed in the tail. Killen and coworkers (2010) concluded that a 'hot' process, impact vaporization, is responsible, and that a similar mechanism could be responsible for the presence of another refractory element, calcium. The observed magnesium tail is consistent with 5–8% regolith abundance if 30% of the impact vaporized Mg remains as MgO, and half of the resulting vapor condenses. Researchers observed that although global sputtering doesn't appear to play a major role, locally sputtered

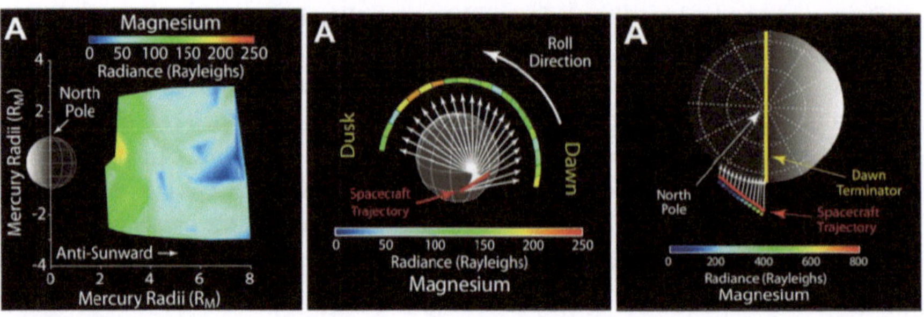

Fig. 4.6 Observations of magnesium taken during first flyby. *Left*: Compilation of illuminated portion of the exosphere. Note northern latitude feature closest to planet. *Middle*: Observations of the fantail role dawn to dusk. Note increase in emission close to dusk. *Right*: Near dawn terminator crossing in equatorial plane. (Reprinted from Icarus, 209, Killen et al. 2010, Figs. 5a, 6a, 7a, © 2010 with permission of Elsevier)

sources could be larger than an IV source and result in the non-uniform distribution observed. Photon desorption of Mg from MgO could be contributing Mg to the dayside if another process (such as impact vaporization) deposits MgO.

Sarantos and coworkers (2011) suggested a hot ejection process (tens of thousands of K) and a source as cool as 400 K for magnesium. Column abundance models indicate that sputtering during times of southward IMF would generate five times more 'hot' atoms than measurements indicate. Sarantos and coworkers (2011) thus propose rapid dissociation of exospheric MgO as the main source of neutral Mg. Sputtering and uniform impact-driven release models can reproduce the observed magnesium tail, but predict only half of the magnesium observed near the terminator. Although the sharp increase in Mg emission near the dawn terminator could be generated by the impact vapor from a single meteoroid, the effect would be more 'hemispheric,' so it was thought that lower energy processes acting upon a reservoir of volatile magnesium a more likely explanation.

Vervack and coworkers (2010) looked at the relationship between neutral and ionized species distributions. In the case of calcium, the neutral species shows a more uniform distribution, making the direct production of localized Ca^+ from Ca unlikely. Vervack and coworkers (2010) propose a mechanism similar to the one proposed for Na^+ formation. Ca^+, with its short ionization lifetime, forms close to the surface, and is picked up through the process of magnetospheric convection followed by centrifugal acceleration across the magnetotail. The pickup velocity is independent of mass, but dependent on acceleration due to magnetic field convection. As indicated in the last section, low energy processes (PSD) with localized high-energy processes (i.e., ion sputtering), are used to explain enhanced high latitude Na emissions over both poles. N/S asymmetries in Ca and Mg could reflect variations in regolith abundances. The more refractory atoms gradually decline from poles to tail, and have high scale height, implying more energetic source production. The magnesium abundance profile is strikingly higher in the north, suggesting an additional source from sputtering from more magnesium-rich regolith (Vervack et al. 2010; Burger et al. 2012). The dawnside peak in Ca fantail observations under variable magnetospheric conditions, implying energetic release mechanisms, is difficult to explain when no such feature occurred in either Mg or Na observations (Vervack et al. 2010; Burger et al. 2012). The consensus is that source and sink processes for each exospheric species are distinctly different, and complex, often involving multiple mechanisms, some still not completely understood.

4.6 Dust in the Exosphere

Interplanetary dust, typically thought of in terms of the micrometeoroid component ($<10^{-6}$ g), is pervasive and by far the largest source of bombardment, impacting directly and unchanged in terms of kinetic energy onto bodies with no atmosphere. Impact energy is then partitioned into surface fracturing shockwave creation, melting, and the release of vapor (Grotheer and Livi 2014), the latter being considered here. So energetic is this process that even refractory elements are vaporized, resulting in an exosphere that 'samples' regolith composition. Although larger (than micrometeoroid) impacts release more kinetic energy and increase atmospheric density, they are far less frequent and influential. For example, impact rates are 44/day vs. 644/day for 10 g and 2.1×10^{-4} g particles, respectively. Grotheer and Livi (2014) modeled vapor production rates as a function of differential mass and velocity distribution (Fig. 4.7), utilizing previous work on dust mass distribution from HEOS 2, Pioneers 8 and 9 (Grun et al. 1985), differential velocity distribution models (Zook 1975), estimated impact rates as a function of mass and velocity (Cintala 1992), differential meteoroid velocity distribution as a function of heliocentric distance (Morgan et al. 1988), and theoretical rate of impact-induced vapor production (Berezhnoy and Klumov 2008). Although they considered particles with

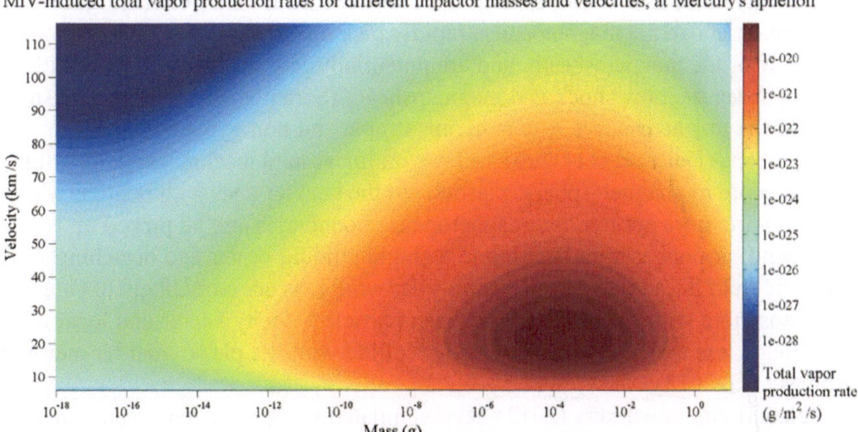

Fig. 4.7 Vapor production and velocity distribution for micrometeoroid of mas 2×10^{-4} g. (Reprinted from Icarus, 227, Grotheer and Livi 2014, Fig. 4, © 2014 with permission of Elsevier)

masses ranging from 10^{-18} to 10^{1} g in this study, Grotheer and Livi (2014) indicated that 90 % of the vapor is produced by particles with masses ranging from 4×10^{-7} to 8×10^{-2} g. The largest amount of mass is contributed by particles 2×10^{-4} g. Particles $< 10^{-9}$ in mass are not subject to solar gravity or the Poynting-Robertson effect, but pushed beyond the Solar System by solar radiation pressure (Borin et al. 2009). Grotheer and Livi (2014) assumed all impacts were normal to the surface in their study. Impacts at more acute angles would result in more abundant, but lower temperature, vapor (Schultz 1996).

4.7 Exosphere-Magnetosphere Interactions

Mercury's magnetic field normally inhibits solar wind access to the dayside equatorial regions, forming a magnetosphere, where magnetic field lines are oriented perpendicular to the velocity vector of the incoming solar wind (Russell 1987). The resulting magnetosphere is compressed by ram pressure on the dayside. At high latitudes, the solar wind interaction with the magnetic field forms funnel-shaped indentations in the magnetopause; these 'cusps' capture some of this flux and guide it to the surface. On the nightside, the magnetic field lines are pushed back and elongated to form the magnetic tail. The magnetosphere is surrounded by ions that originate in the magnetosphere. The distribution of ions in the exosphere and on the surface results directly from these interactions.

The compositional variability and density distribution of the plasma results in interactions of neutral atoms and plasma through processes involving the surface, exosphere, solar wind, solar photons, and magnetosphere. Neutral species, such as Na, Mg, or Ca, may lose an ion through processes such as ion sputtering or PSD or photo-dissociation and become 'pickup' ions—they are picked up as charged particles by Mercury's magnetic field. Each process generates a different temperature, energy, and, as a result, spatial distribution. Pickup ions originating through dissociation above the magnetopause rapidly pick up speed and become part of the bulk flow of the magnetosheath, and are potentially responsible for the 'boundary' layer between the bow shock and magnetosheath (Sarantos et al. 2009). The small scale of the magnetosphere results in anisotropic and non-gyrotropic distributions of heavy ions (Delcourt et al. 2003). Ions gyrating around local magnetic field lines are picked up by magnetospheric plasma, but the heavier ones with relatively large gyroradii are more apt to cross magnetopause boundaries and be picked up by the solar wind, no longer gyrating around a single guiding center and detaching from magnetic field lines; planetary ions and solar wind ions are more likely to enter the magnetosphere with pitch angles (between particle velocity vector and local magnetic field) less than the loss cone angle, collide with the planet, and be removed (Raines et al. 2013).

Delcourt and coworkers (2012) have simulated exospheric ion escape, investigating transport mechanisms for ions in the planet's low altitude magnetosphere and demonstrating centrifugal acceleration as an important mechanism for supplying plasma to the magnetosphere. They demonstrated that centrifugal effects resulting from curvature in E × B drift paths could greatly energize particles in parallel directions, due to the small spatial scales of Mercury's exosphere. As a result, even lower energy ions with initial speeds less than escape velocity, such as those produced by thermal desorption, can escape into the magnetosphere. The magnetospheric convection (movement) rate generally controls the escape route, which becomes narrower as the convection rate increases.

Zurbuchen and coworkers (2011) reported on the spatial distributions of ions derived from MESSENGER flyby FIPS and EPPS data. Photoionization of exospheric neutral atoms taking place regardless of magnetic boundaries is a major source of the ions being measured by FIPS. Once formed, ions are subjected to magnetic and electrical forces, and their distribution, similar to the neutrals at first, becomes progressively different. These researchers consistently observed flux enhancements (1) on the dayside at high northern latitudes in the magnetospheric cusp region, (2) on the nightside at the equator associated with the central plasma sheet, and (3) near and spanning the magnetopause. O^+ and Na^+ ions exhibit distinct maxima in the northern magnetic-cusp region, indicating that the high latitude regions are important source regions for the ionized exosphere, through solar wind sputtering.

Zurbuchen and coworkers (2011) indicated that Na^+ pressure could be a substantial fraction of the total proton pressure in the nightside equatorial region. He^+ is more evenly distributed, indicative of uniformly distributed evaporation from the helium (He) saturated surface. The observed He^+ flux is largely produced locally, with a contribution from interstellar gas, ionized near the Sun and swept along with the solar wind. Peak Na and O fluxes occur at the northern polar regions, associated with surface sputtering near the polar region cusps. Ionized components associated with the global neutral exosphere would be distributed more evenly. Zurbuchen and coworkers (2011) observed comparable fluxes at dawn and dusk terminators.

Raines and coworkers (2013) combined FIPS flyby data (Zurbuchen et al. 2011) with 2 years of orbital data to characterize more comprehensively the composition and distribution of the He^{2+}, and the Na^+, O^+, and He^+ groups in the plasma environment around Mercury. Due to the local nature of FIPS ion density measurements and the variable environmental conditions, data were summed across orbits to derive mean ion densities as a function of altitude, time of day, and time of year (Table 4.3; Raines et al. 2013). Note smaller ratios to Na^+ for O^+ and He^+ groups and larger ratio to Na^+ for He^{2+} for orbital densities over highly elliptical orbits, indicating greater overall abundance of Na^+ relative to O^+ and He^+ particularly closer to the planet, but greater overall abundance of He^{2+} that originates in the solar wind.

Excluding protons, which were not included in this study and accounted for 97 % of the ions, He^{2+} ions were most abundant, but the Na^+ group was most abundant in the northern cusp. Subsolar densities decreased for all ions, but especially for Na^+ ions that were half as abundant. Ions generated at least partially from the regolith ('planetary ions') showed the same pattern in distribution as a function of altitude, time of day, and time of year, but on different scales, and exhibited greater abundance closer to the planet, as would be expected. The pre-midnight enhancement of Na^+ is consistent with previously predicted non-adiabatic motion (Fig. 4.8).

Table 4.3 Densities, peak altitude, and e-folding density for various ion groups

Species	Orbital density (n_{obs}/ cm^3), Ratio to Na+	M1+M2 density (n_{obs}/ cm^3), Ratio to Na+	Dawn Altitude (km), e-folding (km)	Subsolar Altitude (km), e-folding (km)	Dusk Altitude (km), e-folding (km)
Na^+ group	5.1×10^{-3}, 1.0	3.3×10^{-3}, 1.0	670, 800	460, 1100	580, 2500
O^+ group	8.0×10^{-4}, 0.16	1.0×10^{-3}, 0.31	1330, 800	760, 1100	1400, 2100
He^+ group	3.4×10^{-4}, 0.07	2.2×10^{-3}, 0.68	4040, 500	1140, 1300	2790, 70
He^{2+}	3.9×10^{-2}, 7.7	1.0×10^{-3}, 0.31			

Prominent peaks in Na^+ density at TAA's of 110, 150, and 330 (Raines et al. 2013) are consistent with exospheric models (Leblanc and Johnson 2010; Wang and Ip 2011) and observations. The O^+ group exhibits a similar trend, at lower densities, and with less prominent features at 110 and 150° TAAs. In comparison, He^+ is nearly constant. Figure 4.8 illustrates patterns in altitude and local time distributions. Na^+ ions show enhancements at northern latitudes associated with the magnetic field cusp, and between dusk and midnight across all latitudes, which is consistent with neutral and ion transport models and observations. O^+ displays the same pattern but on a smaller scale. Both of these ion groups are escaping neutrals, which originated in the regolith, and are photoionized as they escape. The He^{2+} enhancements are associated with the magnetosheath, indicating its origin as a solar wind component subject to charge exchange as it enters the magnetosphere. He^+ weakly exhibits features of $He2^{2+}$, and Na^+, indicating origin both as an external (solar wind) component and internal (solar wind He-implanted in the regolith) component.

Parameters associated with e-folding (factor of e change in density) (Table 4.3) indicate that Na^+, with its more compact distribution, higher near-surface density, and features at dawn and dusk, originates from Na neutral regolith components. O^+ shows a similar, though less pronounced, pattern. He^+ does not exhibit this pattern. Variations in e-folding heights are probably related to the ratio of the ion gyroradius and neutral scale height (Hartle et al. 2011). Planetary ions form upstream from neutrals and then collect at the subsolar magnetopause, where the plasma flow has a low velocity. They then gyrate across and out of the magnetosphere due to large gyroradii (Spreiter et al. 1966).

ULF waves in Mercury's magnetosphere were first described by Russell (1989). MESSENGER detected ULF transverse (Alfven, particle motion parallel to magnetic field gradient) waves in Mercury's inner magnetosphere, with wave frequencies near H^+, He^+, and He^{2+} gyrofrequencies, along with resonance waves between these frequencies at the same location. Ion-ion hybrid simulations made by Kim and coworkers (2013) are consistent with observations. Observed frequencies are affected by ion density ratios rather than absolute densities, and thus can be used to estimate abundances of heavier ions, which occur in far lower abundances and are less easily observed directly.

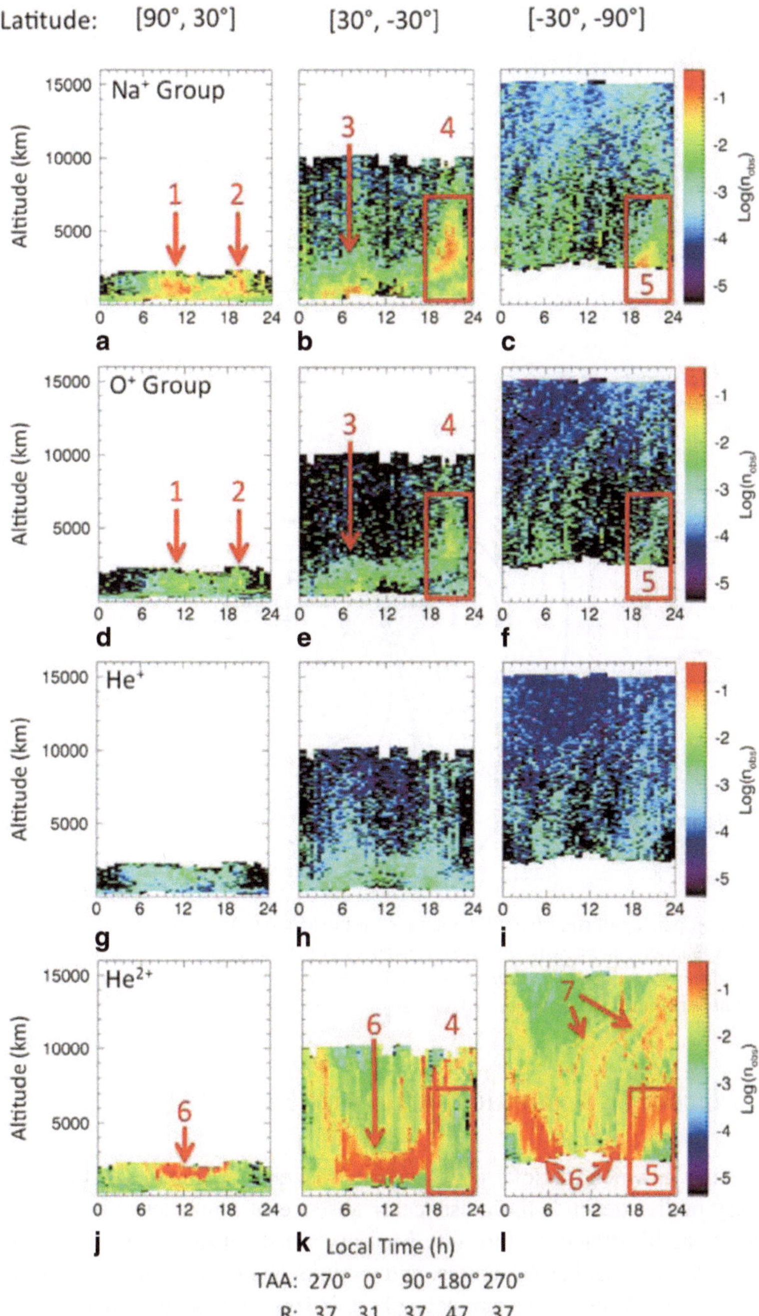

Fig. 4.8 Average observed ion density for all species as a function of altitude and local time in three latitude ranges. (Raines et al. 2013, JGR planets, Fig. 3, © 2013 American Geophysical Union)

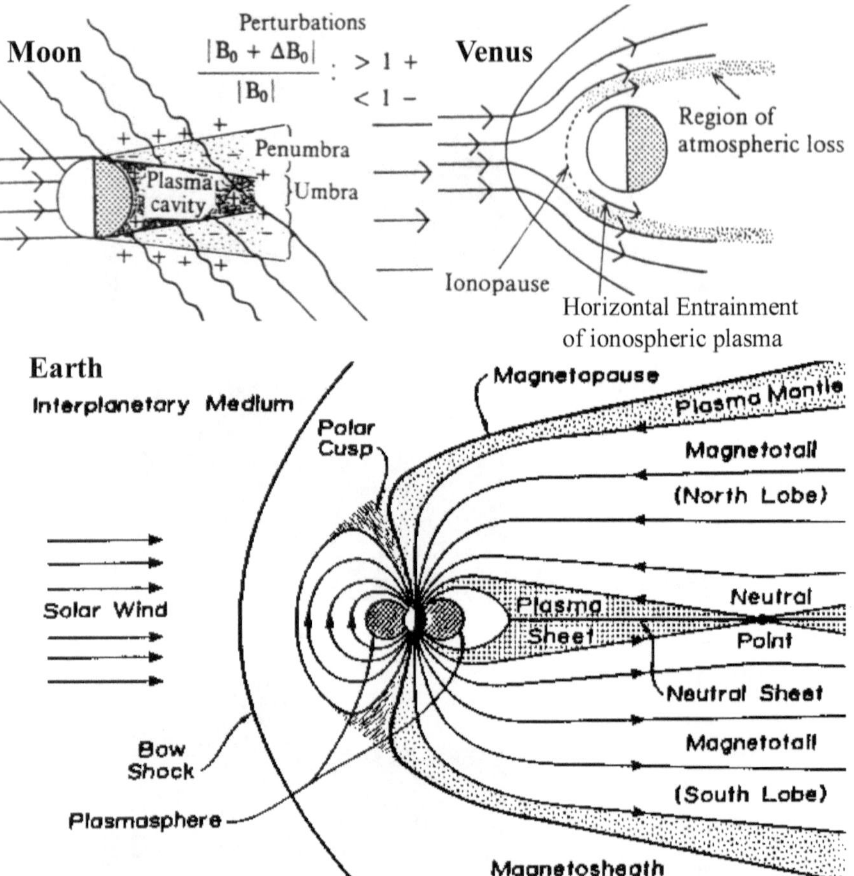

Fig. 4.9 Comparison of solar wind interactions for the Moon (no magnetosphere or atmosphere), Venus (no magnetosphere with atmosphere), and Earth (magnetosphere and atmosphere) *Top*: Modified from Lepping 1986. *Bottom*: Russell 1987, Fig. 6, in *The solar wind and the earth.* (Copyright Terra Scientific Publishing Company)

4.8 Mercury's Dynamic Magnetosphere

The manner in which the solar wind and interplanetary magnetic field interact with planetary bodies depends the presence or absence of a magnetosphere and an atmosphere, as illustrated by Fig. 4.9. Mercury's magnetosphere is closely aligned with its rotation axis, has structures, and exhibits phenomena analogous to Earth's magnetosphere. Although its magnetosphere is 100 times weaker than Earth's, it is usually strong enough to stave off the solar wind (Clark 2007). Mercury's lack of an ionosphere, as well as its exposure to an order of magnitude greater solar wind flux, and five times greater IMF, combined with its far weaker magnetic moment (roughly three orders of magnitude less than Earth's) contribute to its far more rapid changes and variability in response to external events. Its far greater eccentricity contribute to more pronounced differences at perihelion and aphelion.

Despite a lack of instruments designed to study a magnetosphere, observations made by cosmic ray telescopes during the three flybys of *Mariner 10* led to the discovery of a bow shock diverting the incoming plasma (ion foreshock) at about 1 radius, as well as evidence for a magnetopause (magnetosphere outer boundary), magnetosheath (between bow shock and magnetopause), magnetotail, and plasma sheet with electron energy spectrum analogous to Earth's (Winslow et al. 2013). Very intense energetic particle events and field perturbations occurred on rapid time scales. If Mercury's magnetosphere were scaled (by a factor of 8) to occupy the same volume as Earth's magnetosphere, Mercury would occupy a large fraction of Earth's inner magnetosphere, including the plasmasphere and most intense Van Allen belts (Clark 2007). *Mariner 10* provided 0.6 s time resolution measurements of energetic electrons (>35 KeV), and 0.04 s resolution measurements of Mercury's magnetic field. These combined observations indicated that substorms can occur on scales of minutes, a scale more than an order of magnitude faster than Earth's (Christon 1987).

MESSENGER traversed Mercury's magnetosphere from 1 to several Mercury radii, during times of northward, southward, and variable IMF conditions, providing observations of changes in Mercury's magnetosphere structure and solar wind interactions during the range of IMF conditions (Slavin et al. 2008, 2009, 2012a). Mercury's smaller and weaker magnetosphere is more sensitive to IMF direction and more dominated by reconnection than Earth's (Slavin et al. 2009). 3D models created using a global hybrid simulation show strikingly different structure and activity for these two orientations (Travnicek et al. 2010) (Fig. 4.10). The same structures are seen in both cases, including cusps and a closed ion ring belt, but the locations of these features is shifted. Due to the southward orientation of Mercury's magnetic dipole, the southward IMF creates a smaller volume magnetosphere with cusps at lower latitudes sunward, but a larger volume magnetosphere with thinner plasma sheet and more frequent reconnection, plasmoid formation with traveling compression regions (TCR), and flux transfer events anti-sunward, without unloading of tail lobes (Slavin et al. 2012a). Northward IMF results in a steady, long tail, thick low-latitude boundary layers, instabilities, and boundary waves in tail flanks (Slavin et al. 2012a). When the radial component of the IMF is significant, as it typically is, the quasi-neutral line, which occurs in the middle of the neutral plasma sheet within the magnetotail, is shifted to one of the cusps, allowing flux transfer connections to the surface (Belenkaya et al. 2013).

In all cases, the quasi-parallel magnetosheath generates ion-beam driven large-amplitude oscillations, and the quasi-perpendicular magnetosheath exhibits ion temperature anisotropy-driven wave activity (Travnicek et al. 2010). Anderson and coworkers (2013) studied field fluctuations on scales of 0.2–2, 2–20, and 20–300 s as a function of latitude and local time. They found the least fluctuation on the nightside above 30° latitude and in the southern lobe of the magnetotail, and the most fluctuation near the magnetopause boundaries and cusps. During variable IMF conditions,

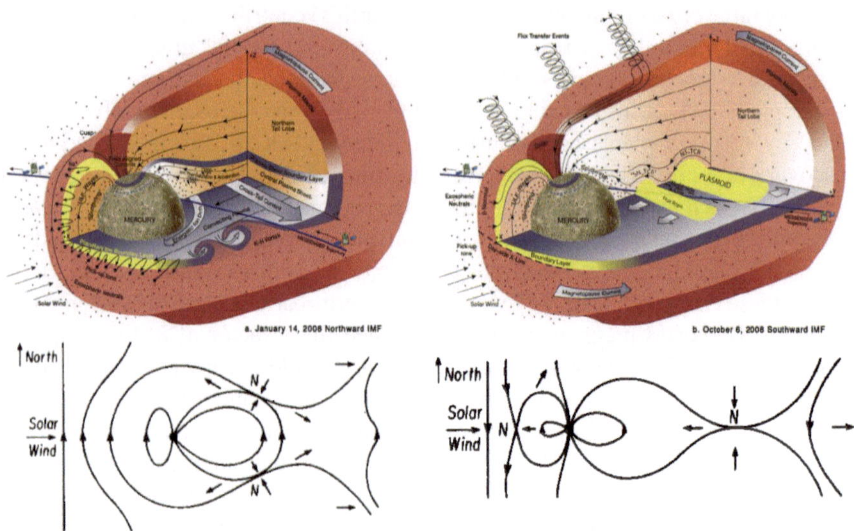

Fig. 4.10 Mercury's magnetic field with IMF oriented northward (*left*) and southward (*right*) as described in text. *Top*: Reprinted from Icarus, 209, Travnicek et al. 2010, Fig. 1, © 2010 with permission from Elsevier. *Bottom*: Russell 1990, The magnetopause, in physics of magnetic flux ropes, Fig. 6, ©1990 American Geophysical Union

the magnetotail is alternately loaded and unloaded with plasma; these events resemble large-scale magnetic field reconfigurations that occur during terrestrial magnetospheric substorms (Slavin et al. 2012a, b).

Magnetic 'cusps' at both poles give the solar wind plasma access to both the surface and magnetosphere of Mercury. During the first 6 months of MESSENGER's orbital mission, Winslow and coworkers (2012) estimated that the northern cusp was 11° in extent at spacecraft altitudes, and estimated, from magnetic pressure deficits, a bombardment rate of approximately 1×10^{24} protons/s over an area of 5×10^{11} m^3. Although plasma pressures at the cusps were estimated to be 40% higher for anti-sunward (outward from cusp) than for sunward (inward to cusp) IMF conditions, which depend on the north/south orientation of the IMF, its influence cannot overcome the inherent north/south asymmetry in Mercury's intrinsic magnetic field (Winslow et al. 2012). As a result the southern cusp, which affects a far larger surface area due to the northward offset in Mercury's magnetic field, would have a flux four times greater than the northern cusp and thus a higher rate of space weathering. Flux was observed to peak at between 65 and 75° latitude in the southern hemisphere, coincident with depressions in magnetic field strength, with an equatorward shift at perihelion (Sarantos 2007).

As indicated by the impact of IMF orientation described above, the influence of external and internal fields can be roughly comparable (Anderson et al. 2010) and thus must both be treated in intrinsic field solutions (Johnson et al. 2012). Due to the weakness of Mercury's intrinsic field, external and internal fields are comparable at

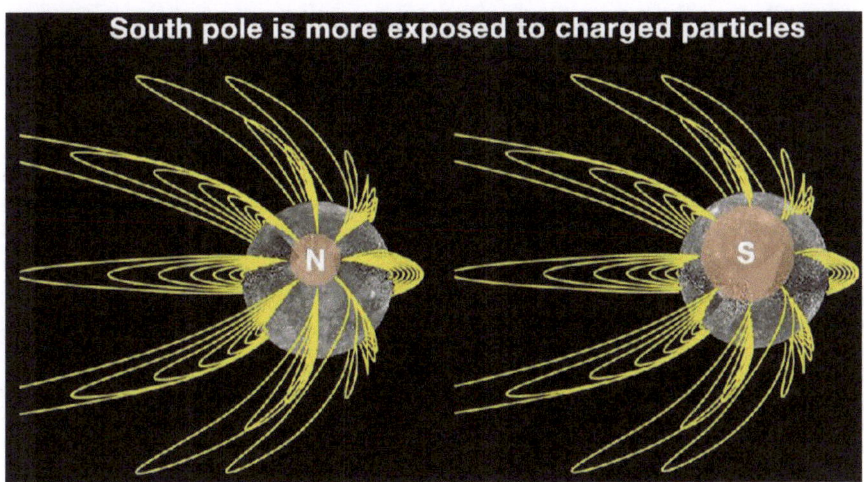

Fig. 4.11 Implications of the offset in Mercury's magnetic field as discussed in the text. (Courtesy of NASA and John Hopkins University Applied Physics Laboratory)

low altitudes, creating difficulty in modeling higher order and degree coefficients to distinguish among likely scenarios. Getting the shape of the magnetosphere, as described below, provides clues to the underlying field structure.

MESSENGER discovered another striking feature of Mercury's magnetosphere: an offset of 479 km (from low altitude crossings) to 486 km (from high altitude crossings with greater uncertainty) northward of the equator centered within 0.8° of the planet's spin axis, indicating offset in Mercury's magnetic dipole (Anderson et al. 2011a, 2012) (Fig. 4.11) with a moment of 190 nT-R_M^3 (Johnson et al. 2012). As a result, the proximity and convergence of magnetic field lines is greater at the north pole, leaving the south pole more exposed to charged particles and space weathering effects (Anderson et al. 2011a). The offset of about 0.19 Mercury radius (R_M) is equivalent to a 0.4 ratio of quadrupole to axial dipole terms in spherical harmonic representation, indicating a large quadrupole term (Anderson et al. 2011a). In addition, Anderson and coworkers (2012) looked at offset in axial alignment to residuals in the degrees 3 (octupole) and 4 spherical harmonic coefficients, and estimated the Gauss coefficient magnitudes for these offsets relative to the dipole to be small, on the order of 4 and 7 %, respectively. These parameters indicate that the magnetic field at the core mantle boundary is weakly dominated by a dipole but has a large quadrupole component. Conventional dynamos, where the top of the convecting layer coincides with the top of the core, would produce such a field with difficulty: models produce a field that is too weak, has too small an offset, or is not dipole dominated. The offset in axial alignment and prominent quadrupole character of the field are consistent with a non-convecting layer in Mercury's fluid outer core above a deeper dynamo. How consistent this is with other observations will be discussed in the next section.

4.9 Mercury's Magnetosphere Constraints on its Magnetic Field

Recent findings, confirming that Mercury has a global magnetic field and indicating that at least a layer of core remains molten (Margot et al. 2007), still favor a dynamo rather than remanent magnetization. A variety of structural models have been proposed to generate a dynamo with a weak dipole compatible with observations, including thin shell dynamos (Stanley et al. 2005; Takahashi and Matushima 2006), deep dynamos enclosed with a stably stratified, electrically conductive layer (Christensen and Wicht 2008; Manglik et al. 2010); magnetopause-generated induction feedback on convecting parts of the core (Glassmeier et al. 2007a, b; Gomez-Perez and Solomon 2010; Heyner et al. 2011); and precipitation of solid iron in radial zones within a liquid outer core (Vilim et al. 2010). Models such as induction feedback and double iron snow layer, that are dominated by odd harmonics, are not compatible with the large quadrupole/dipole ratio, and low upper bounds for third and fourth degree harmonics. The non-convecting layer above a deep dynamo, a variation on one Stevenson (1982) proposed for Saturn, works best. Recent study of the gravity field and mass distribution (Smith et al. 2012) suggests the solid outer shell is too dense to be composed solely of silicates, but may contain an iron-rich lower layer, in the form of solid Fe-S at the top of the core (Hauck et al. 2013), as discussed in Chap. 2. Such a layer could also attenuate the magnetic field structure at high degree and order through its electrical conductivity. The data is also consistent with a stably stratified top layer in a liquid core with the dynamo occurring in a convectively unstable lower portion of the fluid core.

Hiremath (2012) modeled Mercury's internal magnetic field structure as a solution to the magnetic diffusion equation using observations of its magnetospheric environment as a constraint. The resulting structural model was consistent with magnetohydrodynamic (MHD) diffusion in the liquid core combined with dipole and quadrupole structures embedded in the mantle and magnetosphere, and constrained the core radius to 2000 km. Large oscillations, or eigenmode values, obtained from the model indicate that the field is billions of years old and suggest it may have inherited its structure, if the temperature remained above the Curie point through the late heavy bombardment.

4.10 Details of the Solar Wind Impact on Magnetospheric Structures

MESSENGER magnetometer measurements were used to establish 'average' shape and location of the bow shock once the orbital mission began (Winslow et al. 2013). The bow shock is a fast magnetosonic (longitudinal, particle propagation perpendicular to magnetic field gradient) standing wave that diverts the solar wind around the magnetospheric cavity. Its distance and shape will depend on the solar wind

Alfven mach number (ratio of solar wind speed to speed of sound) and to a lesser extent on IMF direction. For average solar wind ram pressure (14.3 nPa), the Alfven mach number is 6.6, the average standoff distance is 1.45 R_M (Mercury radius), and the flaring parameter is 0.5, indicating the magnetosphere is closed on the nightside (>0.5 would indicate an open magnetosphere). The best fit to the shape of the bow shock is a hyperboloid with a radius of 1.96 R_M, varying from 2.29 to 1.89, and an eccentricity of 1.02. Under these conditions, the magnetopause hyperbolic radius varies from 1.55 to 1.35 R_M. The magnetotail surface was found to be not parabolic like Earth's, but nearly cylindrical with a radius of 2.7 R_M at 3 R_M. (Winslow et al. 2012, 2013; Shue et al. 1997), which is consistent with an observed weak dipole with an offset. This shape may also imply that magnetic flux has a short residence time in the tail, implying rapid convection from the tail to the dayside. The standoff distance is correlated with the Alfven mach number for both Earth and Mercury. However, the flaring of Earth's bow shock is also correlated with mach number, and the flaring of Mercury's bow shock does not change. The expansions and contractions of the magnetopause, as reconnection occurs in response to external events, takes place on scales of 1 h for Earth, but only minutes for Mercury (Slavin et al. 2009). Sundberg and coworkers (2013) observed nearly periodic conditions, indicative of ULF waves, in the bow shock under steady solar wind conditions, indicating subsolar cyclic reformation of the shock front as the dominant process, rather than the more complex interactions of shorter but larger-amplitude events and structures observed for Earth under roughly similar conditions.

Le and coworkers (2013) studied the region upstream of the bow shock (foreshock region). They found ULF waves with typical frequencies of 2 Hz analogous to 1 Hz whistler waves in Earth's foreshock region. They observed short electron bursts imposed on whistler waves that were not observed by *Mariner 10.* They also observed magnetosonic (longitudinal, particle motion perpendicular to magnetic field gradient) waves with frequencies of <1 Hz generated by backstreaming ions. These occurred only sporadically due to the relatively weak and small scale of the bow shock region, precluding time for wave growth.

MESSENGER detected a double current sheet, indicated by decreased magnetic field amplitude, at the dayside magnetosphere's inner boundary, just inside the magnetopause, with thicknesses of 1000 and 1400 km regardless of the level of magnetic reconnection activity (Anderson et al. 2011b). The width of the inner edge is on the scale of the proton gyro-radius. This feature is consistent with the presence of an outward plasma pressure gradient perpendicular to the magnetic field, and heavy ions are thought to contribute substantially to this pressure. Hybrid (kinetic/fluid) models (Mueller et al. 2012) based on these observations indicated that the layer most likely formed from protons drifting westward from the cusp, which would generate the observed strong pre-noon/post-noon asymmetry regardless of IMF orientation. The outer sheet is the 'classical' magnetopause, whereas the inner sheet is sustained by a diamagnetic current originating from trapped protons traveling along closed field lines between the poles. The combination of the boundary layer current and the dipole generates a force that prevents the protons from entering the inner magnetosphere, enhancing the outward proton pressure gradient at the inner edge of the boundary. Furthermore, Gershman and coworkers (2013) observed what they called a 'plasma depletion layer' of about 300 km forming above the subsolar magnetopause as a result of pile up, and decreased plasma beta (ratio of plasma thermal pressure to magnetic pressure) of magnetic flux draped around the magnetosphere.

Sundberg and coworkers (2012b) studied the impact of dipolarization (change from elongated magnetosphere to magnetic dipole shape at the onset of solar activity-induced disturbances) on the magnetotail. They observed rapid increase in magnetic field amplitude followed by slower decrease. Such conditions result in non-adiabatic heating of plasma sheet protons, and formation of flux rope and traveling compression region-like structures in the tail. The short lifetime of dipolarization events results from the lack of a steady field aligned current system. When the IMF is primarily southward in orientation, dipolarization events are more Earth-like, and include explosive nightside reconnection and high speed, low density plasma streams in the plasma sheet called 'bursty bulk flows' (Li et al. 2011).

4.11 Plasma and Solar Wind Interactions

Observations of plasma activity and interactions of electrons and ions with the solar wind, surface, and exosphere are made by MESSENGER FIPS and magnetometer data, correlated with the MASCS UV spectrometer data, as described above. Clusters of low to moderate energetic electron (tens to hundreds of KeV) bursts (short, rapidly rising spikes in plasma activity) are frequent occurrences in Mercury's magnetosphere, with the largest bursts, similar in intensity, spectral shape, and pitch angle, occurring at high northern latitudes or near midnight, and the lowest frequency of bursts occurring at dawn and dusk (Ho et al. 2011). Bursts typically have an energetic cutoff around 100 KeV (Ho et al. 2011). DiBraccio and coworkers (2013) have used observations made at magnetopause crossings at orbital periapsis in the subsolar region to characterize energetic electron events in a variety of ways. Energetic electron events are typically a few seconds in duration, have a magnetic field component of 20nT, and a normal magnetic field to total interior (inside magnetopause) field ratio of 0.15, a ratio that is a factor of three larger than such a ratio for Earth. The difference between the ratios for Mercury and Earth are correlated with the Alfven speeds of the solar wind, which varies as a function of solar distance. The reconnection rate at Mercury's magnetopause is relatively high, regardless of IMF orientation, because the rate varies inversely with the solar plasma beta, which is lower in the inner heliosphere.

Walsh and coworkers (2013) modeled the capture or loss of energetic particle populations in the magnetosphere. Many particles are lost due to shadowing by the magnetopause. A relatively small proportion of particles, those with relatively high pitch angles, remain trapped for far shorter periods, several orbits, than analogous particles in Earth's magnetosphere (weeks to months). Particle orbits are typically found above high latitudes on the dayside and equatorial latitudes on the night side. Due to their large gyro-radii, energetic H^+ and Na^+ ions typically are within the loss cone and typically collide with Mercury, unable to complete full drift orbits (Walsh et al. 2013). Mercury's offset dipole creates asymmetric loss cones, resulting in greater particle precipitation in the southern hemisphere. Particle collisions with the surface produce space weathering as well as neutrals that contribute to the exosphere.

Slavin and coworkers (2012a) studied the characteristics of an intense cluster of flux transfer events (FTEs), called an 'FTE shower' that occurred in April of 2011 shortly after MESSENGER orbital insertion. They observed 163 FTE's, each lasting 2–3 s and occurring at 8–10 s intervals, in 25 min. Similar events at Earth last 1–2 min and occur at 8 min intervals. Calculated flux rope characteristics include an elliptical cross-section, a mean axial flux of 1.25 MWb, and a mean semimajor axis of 0.15 RM. The northward orientation of the IMF suggested formation just tailward of the southern cusp of the magnetic field, an interpretation confirmed by the model. The typical decrease in magnetic field intensity associated with the FTE shower is thought to result from diamagnetism of new magnetosheath plasma injected into the tail by the FTEs. Such dayside reconnection events replace closed magnetic field lines with pairs of open field lines, one of which is connected to the IMF, that are pushed anti-sunward by the plasma flow in the magnetosheath and that add flux to the northern and southern lobes of the magnetotail. If the IMF had had a southward orientation, reconnection could have occurred either forward of the magnetopause or tailward of either northern or southern cusps, but the high latitude reconnection would have added no new flux to the tail.

Sundberg and coworkers (2012a) frequently observed sawtooth wave patterns in the dayside magnetosphere resulting from turbulence (Kevin Helmholtz instability vortices) generated due to velocity inhomogeneities when the solar wind and magnetospheric particles meet at the magnetopause, and plasma is transported into the magnetosphere. These patterns occurred asymmetrically post-noon and when the IMF was oriented northward, as they do for Earth's magnetosphere. However, the data imply far greater instability growth rates in Mercury's magnetopause than in Earth's, which they attribute to limited energy dissipation in Mercury's highly non-conducting regolith (Miura 1996).

Schriver and coworkers (2011) used global hybrid simulations to model interactions between Mercury's surface, surface-generated exospheric components, and the magnetosphere for conditions observed during the first two MESSENGER flybys. They considered the effectiveness of electron stimulated desorption (ESD) in contributing to electron mobility and precipitation, exosphere generation, and the ion cloud. They achieved qualitative agreement between the model and observed data, in terms of energy and spatial distributions, as well as the lack of high-energy electrons under MESSENGER conditions. Electron ESD production rate was found to be comparable with predicted ion sputtering yields.

IMF orientation, northward during the first flyby and southward during the second flyby, influenced electron trajectories and entry, exit, and reconnection locations (Schriver et al. 2011). These occurred on the dayside at high latitudes for the northward IM, and at the dayside equator and nightside magnetotail for southward IMF; however, the relatively large radial component for the southward IMF resulted in some electron entry north of the equator as well. The highest energy electrons recorded as bursts by *Mariner 10* (>35 KeV) were not observed, possibly due to 'quiet' solar conditions (Schriver et al. 2011). The highest energy (up to 25 KeV) electrons observed become magnetically trapped, whereas those with energies of hundreds of eV's precipitated on the surface.

Reconnection in Earth's near magnetotail, resulting from a combination of non-adiabatic (demagnetization) and adiabatic motion, accelerates less energetic electrons (tens of eV's) to the KeV range in the plasma sheet, resulting in field-aligned beams (Schriver et al. 1998). Non-adiabatic motion is thought to play a larger role in electron acceleration, transport, and reconnection in Mercury's weaker field (Slavin et al. 2009, 2010). Electrons are energized in two steps, analogous to those occurring in Earth's magnetosphere. Acceleration near the reconnection region kicks up the energy to hundreds of KeV's. Electrons are further energized by betatron acceleration (Schriver et al. 2011). According to Schriver and coworkers (2011), electrons launched from upstream in the solar wind are adiabatic and move 'like beads on a string' along IMF lines. About 80 % move more rapidly than solar wind convection and exit quickly without entering the magnetosphere, and 20 % cross the magnetopause boundary, circulate in the magnetosphere, and exit downstream through the magnetotail (15 %) or precipitate on the surface (5 %) (Schriver et al. 2011). Electrons in the magnetotail naturally drift toward dawn due to the radius of curvature and gradient of the B component. Electron precipitation occurs along open field lines occurring in a narrow longitude region at high latitudes near noon for the northward IMF, and in a broader region, due to more spread in pitch and phase angle generating a wider spread in latitude and longitude, at lower latitudes for the southward IMF (Schriver et al. 2011).

References

Anderson, B., Acuna, M., Korth, H., Slavin, J., Uno, H., Johnson, C., Purucker, M., Solomon, S., Raines, J., Zurbuchen, T., GLoeckler, G., McNutt, R.: The magnetic field of Mercury. Sp. Sci. Rev. **152**, 307–339 (2010)

Anderson, B., Johnson, C., Korth, H., Purucker, M., Winslow, R., Slavin, J., Solomon, S., McNutt, R., Raines, J., Zurbuchen, T.: The global magnetic field of Mercury from MESSENGER orbital observations. Science. **333**, 1859–1862 (2011a)

Anderson, B., Slavin, J., Korth, H., Boardsen, S., Zurbuchen, T., Raines, J., Gloeckler, G., McNutt, R., Solomon, S.: The dayside magnetospheric boundary layer at Mercury. Planet. Sp. Sci. **59**, 2037–2050 (2011b)

Anderson, B., Johnson, C., Korth, H., Winslow, R., Borovsky, J., Purucker, M., Slavin, J., Solomon, S., Zuber, M., McNutt, R.: Low-degree structure in Mercury's planetary magnetic field. J. Geophys. Res. Planet. **117**, E00L12 (2012). doi:10.1029/2012JE004159

Anderson, B., Johnson, C., Korth, H.: A magnetic disturbance index for Mercury's magnetic field derived from MESSENGER magnetometer data. Geochem. Geophys. Geosystems. Tech. Brief. **14**(9) 3875–3886 (2013)

Belenkaya, E., Alexeev, I., Slavin, J., Blokhina, M.: Influence of the solar wind magnetic field on the Earth and Mercury magnetospheres in the paraboloidal model. Planet. Sp. Sci. **75**, 46–55 (2013). doi:10.1016/j.pss.2012.10.013

Berezhnoy, A.: Chemistry of impact events on the moon. Icarus. **226**, 205–211 (2013)

Berezhnoy, A., Klumov, B.: Impacts as sources of the exosphere on Mercury. Icarus. **195**, 511–522 (2008)

Berezhnoy, A., Mangano, V., Mura, A., Milillo, A., Orsini, S.: Density distribution of metal-containing species in the exosphere of Mercury after meteoroid impacts. EPSC-DPS 2011 Joint Meet. **6**, EPSC-DPS2011-1793 (2011)

Bida, T., Killen, R., Morgan, T.: Discovery of Ca in the atmosphere of Mercury. Nature. **404**, 159–161 (2000)

Borin, P., Cremonese, G., Marzari, F., Bruno, M., Marchi, M.: Statistical analysis of micrometeoroids flux on Mercury. Astron. Astrophys. **503**(1) 259–264 (2009)

Burger, M., Killen, R., Vervack, R., Bradley, E., McClintock, W., Sarantos, M., Benna, M., Mouawad, N.: Monte Carlo modeling of sodium in Mercury's exosphere during the first tow MESSENGER flybys. Icarus. **209**, 63–74 (2010)

Burger, M., Killen, R., McClintock, W., Vervack, R., Merkel, A., Sprague, A., Sarantos, M.: Modeling MESSENGER observations of calcium in Mercury's exosphere. J. Geophys. Res. Planet. **117**, E00L11 (2012)

Burger, M., Killen, R., McClintock, W., Merkel, A., Vervack, R., Cassidy, T., Sarantos, M.: Seasonal variations in Mercury's dayside calcium exosphere. Icarus. **238**, 51–58 (2014)

Christensen, U., Wicht, J.: Models of magnetic field generation in partly stable planetary cores: Applications to Mercury and Saturn. Icarus. **196**, 16–34 (2008)

Christon, S.: A comparison of the Mercury and Earth magnetospheres: Electron measurements and substorm time scales. Icarus. **71**, 448–471 (1987)

Cintala, M.: Impact-induced thermal effects in the lunar and Mercurian regoliths. J. Geophys. Res. Planet. **97**(E1) 947–974 (1992)

Clark, P.E.: Mercury's magnetosphere. In: Dynamic Planet: Mercury in the Context of its Environment, pp. 139–184. Springer (2007)

Delcourt, D., Grimaldi, S., Leblanc, F., Berthelier, A., Milillo, A., Mura, A., Orsini, S., Moore, T.: A quantitative model of the planetary Na+ contributions to Mercury's magnetosphere. Ann. Geophys. **21**, 1723–1736 (2003)

Delcourt, D., Seki, K., Terada, N., Moore, T.: Centrifugally stimulated exospheric ion escape at Mercury. Geophys. Res. Lett. **39**, L22105 (2012)

DiBraccio, G., Slavin, J., Boardsen, S., Anderson, B., Korth, H., Zurbuchen, T., Raines, J., Baker, D., McNutt, R., Solomon, S.: MESSENGER observations of magnetopause structure and dynamics at Mercury. J. Geophys. Res.: Sp. Phys. **118**, 997–1008 (2013)

Gershman, D., Slavin, J., Raines, J., Zurbuchen, T., Anderson, B., Korth, H., Baker, D., Solomon, S.: Magnetic flux pileup and plasma depletion in Mercury's subsolar magnetosheath. J. Geophys. Res.: Sp. Phys. **118**, 7181–7199 (2013)

Glassmeier, K., Auster, H., Motschmann, U.: A feedback dynamo generating Mercury's magnetic field. Geophys. Res. Lett. **34**, L22201 (2007a). doi:10.1029/2007GL031662

Glassmeier, K., Grosser, J., Auster, U., Constantinescu, D., Narita, Y., Stellmach, S.: Electromagnetic induction effects and dynamo action in the Hermean system. Sp. Sci. Rev. **132**, 511–527 (2007b)

Gomez-Perez, N., Solomon, S.: Mercury's weak magnetic field: A result of magnetospheric feedback? Geopgys. Res. Lett. **37**, L20204 (2010). doi:10.1029/2010GL044533

Grotheer, E., Livi, S.: Small meteoroids' major contribution to Mercury's exosphere. Icarus. **227**, 1–7 (2014)

Grun, E., Zook, H., Fechtig, H., Glese, R.: Collisional balance of the meteoritic complex. Icarus, **62**(2) 244–272 (1985)

Hartle, R., Sarantos, M., Sittler, E.: Pickup ion distributions from three-dimensional neutral exospheres. J. Geophys. Res. **116**, A10101 (2011)

Hauck, S., Margot, J.-L., Solomon, S., Phillips, R., Johnson, Ca., Lemoine, F., Mazarico, E., McCoy, T., Padovan, S., Peale, S., Perry, M., Smith, D, and Zuber, M.: The curious case of Mercury's internal structure, JGR Planets, **118**, 1204–1220 (2013)

Heyner, D., Wicht, J., Gomez-Perez, N., Schmitt, D., Auster, H., Glassmeier, K.: Evidence from numerical experiments for a feedback dynamo generating Mercury's magnetic field. Science. **334**, 1690–1693 (2011)

Hiremath, K.: Magnetic field structure of Mercury. Planet. Sp. Sci. **63–64**, 8–14 (2012)

Ho, G., Krimigis, S., Gold, R., Baker, D., Slavin, J., Anderson, B., Korth, H., Starr, R., Lawrence, D., McNutt, R., Solomon, S.: MESSENGER observations of transient bursts of energetic electrons in Mercury's magnetosphere. Science. **333**, 1865–1868 (2011)

Hunten, D., Morgan, T., Schemansky, D.: The Mercury atmosphere. In: Vilas, M., Chapman, C., Matthews, M. (eds.) Mercury, pp. 562–612. U Arizona Press, Tucson (1988)

Ip, W.: On solar radiation-driven surface transport of sodium atoms at Mercury. Astrophys. J. **356**, 675–681 (1990)

Janches, D., Heinselman, C., Chau, J., Chandran, A., Woodman, R.: Modeling the global micrometeor input function in the upper atmosphere observed by high power and large aperture radars. J. Geophys. Res. Planet. **111**, A07317 (2006)

Johnson, C., Purucker, M., Korth, H., Anderson, B., Winslow, R., Al Asad, M., Slavin, J., Alexeev, I., Phillips, R., Zuber, M., Solomon, S.: MESSENGER observations of Mercury's magnetic field structure. J. Geophys. Res. Planet. **117**, E00L14 (2012). doi:10.1029/2012JE004217

Kameda, S., Yoshikawa, I., Kagitani, M., Okano, S.: Interplanetary dust distribution and temporal variability of Mercury's atmospheric Na. Geophys. Res. Lett. **36**, L15201 (2009)

Killen, R., Ip, W.: The surface bounded atmospheres of Mercury and the Moon. Rev. Geophys. **37**(3) 361–406 (1999)

Killen, R., Sarantos, M.: Source rates and ion recycling rates for Na and K in Mercury's atmosphere. Icarus. **171**, 1–19 (2004)

Killen, R., Bida, T., Morgan, T.: The calcium exosphere of Mercury. Icarus. **173**, 300–311 (2005)

Killen, R., Potter, A., Vervack, R., Bradley, E., McClintock, W., Anderson, C., Burger, M.: Observations of metallic species in Mercury's exosphere. Icarus. **209**, 75–87 (2010)

Killen, R., Hahn, J.: Source of Mercury's calcium exosphere, Icarus (in press)

Kim, E., Johnson, J., Lee, D., Pyo, Y.: Field-line resonance structures in Mercury's multi-ion magnetosphere, Earth Plan. Space, **65**, 447–451 (2013)

Le, G., Chi, P., Blanco-Cano, X., Boardsen, S., Slavin, J., Anderson, B., Korth, H.: Upstream ultra-low frequency waves in Mercury's foreshock region: MESSENGER magnetic field observations. J. Geopys. Res.: Sp. Phys. **118**, 2809–2823 (2013)

Leblanc, F., Doressoundiram, A.: Mercury Exosphere, II: The sodium/potassium ratio. Icarus. **211**, 10–20 (2010)

Leblanc, F., Johnson, R.: Mercury Exosphere, 1.: Global circulation model of its sodium component. Icarus. **209**, 280–300 (2010)

Leblanc, F., Chassefiere, E., Johnson, R., Hunten, D., Kallio, E., Delcourt, D., Killen, R., Luhmann, J., Potter, A., Jambon, A., Cremonese, G., Mendillo, M., Yan, N., Sprague, A.: Mercury's exosphere: Origins and relations to its magnetosphere and surface. Planet. Sp. Sci. **55**, 1069–1092 (2007)

Lepping, R.: Magnetic configurations of planetary obstacles. Proceedings from comparative study of magnetospheric systems, CNES/Cepadues, A87-20701 07-91 (1986)

Li, S., Angelopoulos, V., Runov, A., Zhou, X., McFadden, J., Larson, D., Bonnell, J., Auster, U.: On the force balance around dipolarization fronts within bursty bulk flows. J. Geophys. Res.: Sp. Phys. **116**, A00135 (2011). doi:10.1029/2010JA015884

Manglik, A., Wicht, J., Christensen, U.: A dynamo model with double diffusive convection for Mercury's core. Earth. Planet. Sci. Lett. **289**, 619–628 (2010)

Marchi, S., Morbidelli, A., Cremonese, G.: Flux of meteoroid impacts on Mercury. Astron. Astrophys. **431**(3), 1123–1127 (2005)

Mangano, V., Milillo, A., Mura, A., Orsini, S., deAgnelis, E., diLellis, A., Wurz, P.: The contribution of impulsive meteoritic impact vaporization to the Hermean exosphere. Planet. Sp. Sci. **55**, 1541–1556 (2007)

Margot, J.-L., Peale, S., Solomon, S., Hauck, S., Ghigo, F., Jurgens, R., Yseboodt, M., Giorgini, J., Padovan, S., Campbell, D.: Mercury's moment of inertia from spin and gravity data. J. Geophys. Res. Planet. **117**, E00L09 (2007). doi:10.1029/2012JE004161

McClintock, W., Vervack, R., Bradley, E., Killen, R., Mouawad, N., Sprague, A., Burger, M., Solomon, S., Izenberg, N.: MESSENGER observations of Mercury's exosphere: Detection of magnesium and distribution of constituents. Science. **324**, 610–613 (2009)

Miura, A.: Kelvin-Helmholtz instability for supersonic shear flow at the magnetospheric boundary. Geophys. Res. Lett. **17**, 749–752 (1996)

Morgan, T., Zook, H., Potter, A.: Impact-driven supply of sodium and potassium to the atmosphere of Mercury, Icarus, **75**, 1, 1560179 (1988)

Mouawad, N., Burger, M., Killen, R., Potter, A., McClintock, W., Vervack, R., Bradley, E., Benna, M., Naidu, S.: Constraints on Mercury's Na exosphere: Combined MESSENGER and ground-based data. Icarus. **211**, 21–36 (2011)

Mueller, J., Simon, S., Wang, Y., Motschmann, U., Heyner, D., Schuele, J., Ip, W., Kleindienst, G., Pringle, G.: Origin of Mercury's double magnetopause: 3D hybrid simulation study with A.I.K.E.F., Icarus, **218**, 666–687 (2012)

Mura, A.: Loss rates and time scales for sodium at Mercury. Planet. Sp. Sci. **63–64**, 2–7 (2012)

Pifko, S., Janches, D., Close, S., Sparks, J., Nakamura, T., Nesvorny, D.: The meteoroid input function and predictions of mid-latitude meteor observations by the MU radar. Icarus. **223**, 444–459 (2013)

Potter, A., Morgan, T.: Discovery of sodium in the atmosphere of Mercury. Science. **229**, 651–653 (1985)

Potter, A., Killen, R., Sarantos, M.: Spatial distribution of sodium on Mercury. Icarus. **181**, 1-12 (2006)

Potter, A., Killen, R., Morgan, T.: Solar radiation acceleration effects on Mercury sodium. Icarus. **186**, 571–580 (2007)

Potter, A., Morgan, T., Killen, R.: Sodium winds on Mercury. Icarus. **204**, 355–367 (2009)

Raines, R., Gershman, D., Zurbuchen, T., Sarantos, M., Slavin, J., Gilbert, J., Korth, H., Anderson, B., Gloeckler, G., Krimigis, S., Baker, D., McNutt, R., Solomon, S.: Distribution and compositional variations of plasma ions in Mercury's space environment: The first three Mercury years of MESSENGER observations. J. Geophys. Res. Planet. **118**, 1604–1619 (2013)

Russell, C.: The magnetosphere. In: Akasofu, I., Kamide, Y. (eds.) The Solar Wind and the Earth, pp. 73–100. Terra Scientific Publishing Company, Tokyo (1987)

Russell, C.: ULF waves in the magnetosphere. Geophys. Res. Lett. **16**(11), 1253–1256 (1989)

Russell, C.: The magnetopause. In: Russell, C., Priest, E., Lee, L. (eds.) Physics of Magnetic Flux Ropes, vol. 58, pp. 439–454. Geophysical Monograph (1990)

Sarantos, M., Killen, R., Kim, D.: Predicting the long-term solar wind ion-sputtering source at Mercury. Planet. Sp. Sci. **55**, 1584–1595 (2007)

Sarantos, M., Slavin, J., Benna, M., Boardsen, S., Killen, R., Schriver, D., Travnicek, P.: Sodium-ion pickup observed above the magnetopause during MESSENGER's first Mercury flyby: Constraints on neutral exospheric models. Geophys. Res. Lett. **36**, L04116 (2009)

Sarantos, M., Killen, R., McClintock, W., Bradley, E., Vervack, R., Benna, M., Slavin, J.: Limits to Mercury's magnesium exosphere from MESSENGER second flyby observations. Planet. Sp. Sci. **59**, 1992–2003 (2011)

Schmidt, C.: Monte Carlo modeling of north-south asymmetries in Mercury's sodium exosphere. J. Geophys. Res.: Sp. Phys. **118**, 4564–4571 (2013)

Schmidt, C., Baumgardner, J., Mendillo, M., Wilson, J.: Escape rates and variability constraints for high-energy sodium sources at Mercury. J. Geophys. Res. Planet. **117**, A03301 (2012)

Schriver, D., Ashour-Abdalla, M., Richard, R.: On the origin of the ion-electron temperature difference in the plasma sheet. J. Geophys. Res.: Sp. Phys. **103**, 14879–14895 (1998)

Schriver, D., Travnicek, P., Ashour-Abdalla, M., Richard, R., Hellinger, P., Slavin, J., Anderson, B., Baker, D., Benna, M., Boardsen, S., Gold, R., Ho, G., Korth, H., Krimigis, S., McClintock, W., McLain, J., Orlando, T., Sarantos, M., Sprague, A., Starr, R.: Electron transport and precipitation at Mercury during the MESSENGER flybys: Implications for electron-stimulated desorption. Planet. Sp. Sci. **59**, 2026–2036 (2011)

Schultz, P.: Effect of impact angle on vaporization. J. Geophys. Res. Planet. **101**(E9), 21117–21136 (1996)

Shue, J., Chao, J., Fu, H., Russell, C., Song, P., Khurana, K., Singer, H.: A new functional form to study the solar wind control of the magnetopause size and shape. J. Geophys. Res.: Sp. Phys. **102**, 9497–9511 (1997)

Slavin, J., Acuna, M., Anderson, B., Baker, D., Benna, M., Gloeckler, G., Gold, R., Ho, G., Killen, R., Korth, H., Krimigis, S., McNutt, R., Raines, J., Schriver, D., Solomon, S., Starr, R., Travnicek, P., Zurbuchen, T.: Mercury's magnetosphere after MESSENGER's first flyby. Science. **321**, 85–89 (2008). doi:10.1126/science.1159040

Slavin, J., Acuna, M., Anderson, B., Baker, D., Benna, M., Boardsen, S., Gloeckler, G., Gold, R., Ho, G., Korth, H., Krimigis, S., McNutt, R., Raines, J., Sarantos, M., Schriver, D., Solomon, S., Travnicek, P., Zurbuchen, T.: MESSENGER observations of magnetic reconnection in Mercury's magnetosphere. Science. **324**, 606–610 (2009)

Slavin, J., Acuna, M., Anderson, B., Baker, D., Benna, M., Boardsen, S., Gloeckler, G., Gold, R., Ho, G., Korth, H., Krimigis, S., McNutt, R., Raines, J., Sarantos, M., Schriver, D., Solomon, S., Travnicek, P., Zurbuchen, T.: MESSENGER observations of extreme loading and unloading of Mercury's magnetic tail. Science. **329**, 665–668 (2010)

Slavin, J., Anderson, B., Baker, D., Benna, M., Boardsen, S., Gold, R., Ho, G., Imber, S., Korth, H., Krimigis, S., McNutt, R., Raines, J., Sarantos, M., Schriver, D., Solomon, S., Travnicek, P., Zurbuchen, T.: MESSENGER and Mariner 10 flyby observations of magnetotail structure and dynamics at Mercury. J. Geophys. Res. Planet. **117**, A01215 (2012a). doi:10.1029/2011JA016900

Slavin, J., Imber, S., Boardsen, S., DiBraccio, G., Sundberg, T., Sarantos, M., Nieves-Chinchilla, T., Azabo, A., Anderson, B., Korth, H., Zurbuchen, T., Raines, J., Johnson, C., Winslow, R., Killen, R., McNutt, R., Solomon, S.: MESSENGER Observations of a flux-transfer-event shower at Mercury. J. Geophys. Res. Planets. **117**, A00M06 (2012b). doi:10.1029/2012JA017926

Smith, D., Zuber, M., Phillips, R., Solomon, S., Hauck, S., Lemoine, F., Mazarico, E., Neumann, G., Peale, S., Margot, J.-L/, Johnson, C., Torrence, M., Perry, M., Rowlands, D., Goossens, S., Head, J., Taylor, A.: Gravity field and internal structure of Mercury from MESSENGER. Science. **336**, 214–217 (2012)

Smyth, W., Marconi, M.: Theoretical overview and modeling of the sodium and potassium atmosphere of Mercury. Astrophys. J. **441**, 839–864 (1995)

Spreiter, J., Summers, A., Alksne, A.: Hydromagnetic flow around the magnetosphere. Planet. Sp. Sci. **14**, 223–253 (1966)

Stanley, S., Bloxham, J., Hutchison, W., Zuber, M.: Thin shell dynamo models consistent with Mercury's weak observed magnetic field. Earth. Planet. Sci. Lett. **234**, 27–38 (2005)

Stevenson, D.: Reducing the non-axis symmetry of a planetary dynamo and an application to Saturn. Geophys. Astrophys. Fluid. Dyn. **21**, 113–127 (1982)

Sundberg, T., Boardsen, S., Slavin, J., Anderson, B., Korth, H., Zurbuchen, T., Raines, J., Solomon, S.: MESSENGER orbital observations of large-amplitude Kelvin-Helmholtz waves at Mercury's magnetopause. J. Geophys. Res. Planet. **117**, A04216 (2012a). doi:10.1029/2011JA017268

Sundberg, T., Slavin, J., Boardsen, S., Anderson, B., Korth, H., Ho, G., Schriver, D., Uritsky, V., Zurbuchen, T., Raines, J., Baker, D., Krimigis, S., McNutt, R., Solomon, S.: MESSENGER observations of dipolarization events in Mercury's magnetotail. J. Geophys. Res. Planet. **117**, A00M03 (2012b). doi:10.1029/2012JA017756

Sundberg, T., Boardsen, S., Slavin, J., Uritsky, B., Anderson, B., Korth, H., Gershman, D., Raines, J., Zurbuchen, T., Solomon, S.: Cyclic reformation of quasi-parallel bow shock at Mercury: MESSENGER observations. J. Geophys. Res.: Sp. Phys. **118**, 6475–6464 (2013)

Takahashi, F., Matushima, M.: Dipolar and non-dipolar dynamos in thin spherical shell geometry with implications for the magnetic field of Mercury. Geophys. Res. Lett. **33**, L10202 (2006). doi:10.1029/2006GL025792

Travnicek, P., Hellinger, P., Schriver, D., Hercik, D., Anderson, B., Sarantos, M., Slavin, J., Zurbuchen, T.: Mercury's magnetosphere-solar wind interaction for northward and southward interplanetary magnetic field: Hybrid simulation results. Icarus. **209**, 11–22 (2010)

Vervack, R., McClintock, W., Killen, R., Sprague, A., Anderson, B., Burger, M., Bradley, E., Mouawad, N., Solomon, S., Izenberg, N.: Mercury's complex exosphere: Results from MESSENGER's third flyby. Science. **329**, 672–675 (2010)

Vilim, R., Stanley, S., Hauck, S.: Iron snow zones as a mechanism for generating Mercury's weak observed magnetic field. J. Geophys. Res. Planet. **115**, E11003 (2010). doi:10.1029/2009JE003528

Walsh, B., Ryou, A., Sibeck, D., Alexeev, I.: Energetic particle dynamics in Mercury's magnetosphere. J. Geophys. Res.: Sp. Phys. **118**, 1992–1999 (2013)

Wang, Y., Ip, W.: Source dependency of exospheric sodium on Mercury. Icarus. **216**, 387–402 (2011)

Winslow, R., Johnson, C., Anderson, B., Korth, H., Slavin, J., Purucker, M., Solomon, S.: Observations of Mercury's northern cusp region with MESSENGER's magnetometer. Geophys. Res. Lett. **39**, L08112 (2012). doi:10.1029/2012GL051472

Winslow, R., Anderson, B., Johnson, C., Slavin, J., Korth, H., Purucker, M., Baker, D., Solomon, S.: J. Geophys. Res.: Sp. Phys. **118**, 2213–2227 (2013)

Zook, H.: The state of meteoritic material on the moon. Proceedings 6th Lunar Science Conference, vol. **2,** Pergamon Press, 1633–1672 (1975)

Zurbuchen, T., Raines, J., Slavin, J., Gershman, D., Gilbert, J., Gloeckler, G., Anderson, B., Baker, D., Korth, H., Krimigis, S., Sarantos, M., Schriver, D., McNutt, R., Solomon, S.: MESSENGER observations of the spatial distribution of planetary ions near Mercury. Science. **333**, 1862–1865 (2011)